Large Language Models Ops for Finance

A Practical Guide to Infrastructure, Implementation, and Innovation

Brindha Priyadarshini Jeyaraman

Apress®

Large Language Models Ops for Finance: A Practical Guide to Infrastructure, Implementation, and Innovation

Brindha Priyadarshini Jeyaraman
Singapore, Singapore

ISBN-13 (pbk): 979-8-8688-1699-4 ISBN-13 (electronic): 979-8-8688-1700-7
https://doi.org/10.1007/979-8-8688-1700-7

Copyright © 2025 by Brindha Priyadarshini Jeyaraman

This work is subject to copyright. All rights are reserved by the Publisher, whether the whole or part of the material is concerned, specifically the rights of translation, reprinting, reuse of illustrations, recitation, broadcasting, reproduction on microfilms or in any other physical way, and transmission or information storage and retrieval, electronic adaptation, computer software, or by similar or dissimilar methodology now known or hereafter developed.

Trademarked names, logos, and images may appear in this book. Rather than use a trademark symbol with every occurrence of a trademarked name, logo, or image we use the names, logos, and images only in an editorial fashion and to the benefit of the trademark owner, with no intention of infringement of the trademark.

The use in this publication of trade names, trademarks, service marks, and similar terms, even if they are not identified as such, is not to be taken as an expression of opinion as to whether or not they are subject to proprietary rights.

While the advice and information in this book are believed to be true and accurate at the date of publication, neither the authors nor the editors nor the publisher can accept any legal responsibility for any errors or omissions that may be made. The publisher makes no warranty, express or implied, with respect to the material contained herein.

>Managing Director, Apress Media LLC: Welmoed Spahr
>Acquisitions Editor: Celestin Suresh John
>Development Editor: James Markham
>Editorial Assistant: Gryffin Winkler

Cover designed by eStudioCalamar

Cover image by Pexels from Pixabay

Distributed to the book trade worldwide by Springer Science+Business Media New York, 1 New York Plaza, New York, NY 10004. Phone 1-800-SPRINGER, fax (201) 348-4505, e-mail orders-ny@springer-sbm.com, or visit www.springeronline.com. Apress Media, LLC is a Delaware LLC and the sole member (owner) is Springer Science + Business Media Finance Inc (SSBM Finance Inc). SSBM Finance Inc is a **Delaware** corporation.

For information on translations, please e-mail booktranslations@springernature.com; for reprint, paperback, or audio rights, please e-mail bookpermissions@springernature.com.

Apress titles may be purchased in bulk for academic, corporate, or promotional use. eBook versions and licenses are also available for most titles. For more information, reference our Print and eBook Bulk Sales web page at http://www.apress.com/bulk-sales.

Any source code or other supplementary material referenced by the author in this book is available to readers on GitHub. For more detailed information, please visit https://www.apress.com/gp/services/source-code.

If disposing of this product, please recycle the paper

Dedicated to

My beloved parents:

Mr. Jeyaraman

Mrs. Patturani

and

My husband, Suneet

and

My children, Riaan and Riya

Table of Contents

About the Author .... **xiii**

About the Technical Reviewer .... **xv**

Acknowledgments .... **xvii**

Chapter 1: Introduction to Large Language Models in Finance 1

    Large Language Models Overview 1

        Underlying Mechanisms 2

        Key Advancements in LLM Technology 4

        Capabilities and Limitations 5

        LLM Market Growth and Global Adoption 6

        Efficiency Gains in Finance: Reducing Repetitive Tasks 7

        Strategic Benefits of Efficiency Gains 8

    Applications in Finance 9

        Automated Analysis and Insights 9

        Risk Management 10

        Financial Forecasting and Predictive Analysis 12

        Customer Service and Personal Finance Assistance 12

        Hybrid AI Approaches in Finance 13

        Fraud Detection and Regulatory Compliance 14

    Benefits of Using LLMs in Finance 15

        Enhanced Efficiency and Productivity 15

        Improved Accuracy and Reduced Human Error 16

        Advanced Analytical Capabilities 17

        Real-Time Risk Management 18

        Enhanced Client Experience and Personalization 18

        Improved Regulatory Compliance and Reporting 19

TABLE OF CONTENTS

Challenges in Financial Applications 20
- Data Sensitivity and Privacy 20
- Model Interpretability and Transparency 21
- Handling Bias and Ensuring Fairness 22
- Model Drift and Data Volatility 25
- Compliance with Regulatory Standards 26
- Managing High Computational Costs 27

Conclusion 28

Chapter 2: Infrastructure Setup for LLMs 29

Hardware Requirements 29
- Processing Power: GPUs, TPUs, and CPUs 30
- Storage Solutions 31
- Memory Configurations 32
- Optimizing Hardware for Financial LLMs 33

Software Stack 35
- Libraries and Frameworks 35
- Model Management Tools 37
- Deployment and Orchestration Tools 39
- Data Processing and Integration Tools 41

Cloud vs. On-Premises Solutions 42
- Cloud Infrastructure 42
- Challenges of Cloud Infrastructure 44
- On-Premises Infrastructure 44
- Challenges of On-Premises Infrastructure 46
- Hybrid Solutions 47

Best Practices for Infrastructure Setup 48
- Security Guidelines 48
- Redundancy Guidelines 50
- Compliance Guidelines 51
- Best Practices for Financial-Specific Infrastructure 53

Conclusion 55

Chapter 3: Training and Fine-Tuning LLMs 57

Goals and Processes of Training and Fine-Tuning 57
- High-Quality Data: The Foundation of Success 60
- Effective Methodologies for Training and Fine-Tuning 65
- Adaptive Strategies for Domain-Specific Challenges 69

Data Preparation for LLMs 72
- Data Collection 73
- Data Cleaning 74
- Data Augmentation 75
- Dataset Splitting 76
- Data Governance and Documentation 76
- Tools and Frameworks for Data Preparation 77

Training Methodologies 78
- Pretraining: Building a Foundational Model 78
- Distributed Training: Scaling for Large Models 79
- Optimization Algorithms 80
- Regularization Techniques 80
- Checkpointing and Logging 81
- Handling Data and Model Bias 81
- Challenges in Training LLMs 82

Fine-Tuning Techniques 83
- Task-Specific Fine-Tuning 83
- Domain-Adaptive Pretraining (DAPT) 84
- Transfer Learning 85
- Few-Shot and Zero-Shot Learning 85
- Hyperparameter Tuning During Fine-Tuning 86
- Regularization Techniques 86
- Multi-Task Fine-Tuning 87
- Active Fine-Tuning 87
- Handling Domain Shifts 88
- Evaluation During Fine-Tuning 88

Challenges in Training and Fine-Tuning 89
Data-Related Challenges 89
Computational Challenges 90
Domain-Specific Challenges 90
Best Practices for Training and Fine-Tuning 91
Data Preparation 91
Training Best Practices 92
Fine-Tuning Best Practices 93
Evaluation and Monitoring 94
Compliance and Ethics 95
Case Study: Fine-Tuning for Risk Management 96
Context and Problem Statement 96
Objective 96
Methodology 97
Conclusion 101

Chapter 4: Deployment Strategies for LLMs 103
Structure 103
Objectives 104
Introduction 104
Deployment Pipelines 104
Monitoring and Logging 113
Performance Optimization 119
Conclusion 132
Points to Remember 133

Chapter 5: Ensuring Data Privacy and Security 135
Structure 135
Objectives 136
Introduction 137
Data Anonymization Techniques 140
Case Study: Fraud Detection 144

Secure Data Storage 145
    Benefits of Cloud-Based Storage 148
    Risks of Cloud-Based Storage 149
    Best Practices 149

Compliance with Financial Regulations 151
    Understanding Financial Data Regulations 151

Challenges and Best Practices 157
    Key Challenges 158

Conclusion 164

Points to Remember 164

Chapter 6: Integrating LLMs into Financial Systems 167

Structure 167

Objectives 168

API Development and Management 168
    Key Guidelines for API Design 169

Real-Time Data Processing 179
    Importance of Real-Time Data Processing in Finance 179
    Techniques for Real-Time Data Processing 180

Case Studies 186

Conclusion 198

Points to Remember 199

Chapter 7: Monitoring and Maintenance of LLMs 201

Structure 201

Objectives 202

Introduction 203
    Ensuring Long-Term Performance 203
    Adapting to Evolving Data 203
    Compliance with Operational Requirements 204
    Challenges in Monitoring and Maintenance 205
    Examples of Monitoring 205

TABLE OF CONTENTS

- Performance Metrics 206
  - Key Performance Metrics for LLMs 207
  - Real-Time Metric Monitoring 210
  - Best Practices for Performance Monitoring 212
- Regular Updates and Retraining 215
  - Adapting to Changing Data 215
  - Responding to Regulatory Changes 217
  - Addressing Model Drift 217
  - Automating Retraining Pipelines 220
  - Best Practices for Regular Updates and Retraining 222
- Handling Model Drift 223
  - Types of Model Drift 223
  - Statistical Techniques for Drift Analysis 225
  - Case Study: Handling Drift in Fraud Detection 228
- Challenges and Best Practices 229
  - Data Privacy and Security 230
- Future Trends 234
  - Predictive Maintenance 235
  - Self-Healing Systems 236
  - Reinforcement Learning-Driven Adaptability 237
  - The Intersection of Trends 238
  - Preparing for the Future of LLM Maintenance 238
- Conclusion 239
- Key Points 240

Chapter 8: Future Trends in LLM Ops for Finance 243
- Structure 243
- Objectives 244
- Overview of Evolving LLMs in Finance 244
  - Adapting to Regulatory Changes 246

Emerging Technologies 247
    Quantum Computing and Finance 247
    Advancements in Model Architectures 252
    Cloud-Native and Edge LLM Deployments 254

Predictive Maintenance of Models 257
    Proactive Monitoring Techniques 258
    Automated Updates and Retraining Pipelines 261
    Ensuring Long-Term Model Reliability 263

The Future of AI in Finance 265
    Regulatory Standards and Ethical Considerations 265
    Emerging Applications of LLMs in Finance 268

Collaboration Between Institutions and Technology Providers 270
    Building Partnerships to Co-Develop Financial LLMs 270
    The Role of Open-Source Contributions 271
    Future Prospects for Collaborative Ecosystems in Financial AI 272

Conclusion 273

Key Points 273

Index 275

About the Author

 **Brindha Priyadarshini Jeyaraman** is a distinguished leader in AI and data science with over 16 years of experience across machine learning, real-time systems, and cloud-scale architecture. She currently serves as Senior Director and Head of AI Governance at UOB, where she drives responsible AI adoption, oversees governance frameworks for GenAI and agentic systems, and ensures compliance across financial AI deployments in the region.

Previously, Brindha was the Principal Architect for AI, APAC at Google Cloud, leading large-scale AI transformations across sectors including finance, telecommunications, gaming, and consumer AI ecosystems. A recognized expert in MLOps, streaming systems, and temporal knowledge graphs, she has been instrumental in shaping enterprise AI strategies and partner ecosystems across APAC.

Brindha holds a Doctor of Engineering in AI from Singapore Management University, with a specialization in Temporal Knowledge Graphs for Finance, and a master's in Knowledge Engineering from the National University of Singapore. She is the author of three authoritative books on machine learning, streaming analytics, and financial observability and is a passionate advocate for ethical AI, mentoring, and inclusive innovation. Widely regarded as a thought leader in the AI community, Brindha combines deep technical expertise with a mission to make AI trustworthy by design.

Brindha is deeply passionate about mentoring the next generation of AI professionals and championing diversity and inclusion in the tech industry. Brindha is recognized for her innovative approach to solving complex problems and is a leading voice in the AI community.

About the Technical Reviewer

Sonal Raj is an engineer, data scientist, and Python evangelist from India who has carved a niche in the financial services domain. He is a Goldman Sachs and D.E. Shaw alumnus who currently serves as a vice president, leading the Data Management & Research division for a prominent high-frequency trading firm.

Sonal holds dual master's in computer science and business administration and is a former research fellow of the Indian Institute of Science. His areas of research range from image processing and real-time graph computations to electronic trading algorithms. Sonal is the author of the titles *Graph Data Analytics* (BPB, 2024), *The Pythonic Way* (BPB, 2021), and *Neo4j High Performance* (Packt, 2015), among others. During his career, Sonal has been instrumental in designing low-latency trading algorithms, trading strategies, market signal models, and components of electronic trading systems. He is also a community speaker and a Python and data science mentor to young minds in the field.

When not engrossed in reading fiction or playing symphonies, he spends far too much time watching rockets lift off.

He is a loving son and husband and a custodian of his personal library.

Acknowledgments

I want to express my deepest gratitude to my family for their unwavering support and encouragement throughout this book's writing, especially my husband, Suneet, and my children, Riaan and Riya.

I am also grateful to Springer Publications for their guidance and expertise in bringing this book to fruition. It was a long journey of revising this book, with the valuable participation and collaboration of reviewers, technical experts, and editors.

I would also like to acknowledge the valuable contributions of my colleagues and coworkers during many years working in the tech industry, who have taught me so much and provided valuable feedback on my work.

Finally, I would like to thank all the readers who have taken an interest in my book and for their support in making it a reality. Your encouragement has been invaluable.

CHAPTER 1

Introduction to Large Language Models in Finance

This chapter introduces large language models (LLMs) and their transformative potential within the finance industry. It explores how advancements in natural language processing have led to the development of sophisticated models capable of automating complex financial analyses, enhancing risk management, and driving strategic decision-making. By understanding what LLMs are and how they apply to finance, you will gain insight into the opportunities and considerations of using these models in a sector that demands precision, reliability, and ethical responsibility.

It also looks at the current state of LLMs, including recent breakthroughs, and presents a comprehensive overview of their applications in finance. From risk assessment to financial forecasting, the chapter discusses various use cases and the operational benefits LLMs bring to financial services. You'll explore the challenges associated with deploying LLMs in finance, such as data sensitivity, model interpretability, and the ethical considerations that arise when integrating AI into decision-making processes. This introduction sets the foundation for the following chapters, which go deeper into the technical, operational, and regulatory aspects of LLM deployment in financial contexts.

Large Language Models Overview

LLMs represent a significant leap forward in artificial intelligence, enabling computers to understand, generate, and respond to human language with remarkable accuracy. At their core, LLMs are sophisticated statistical models designed to predict the probability

CHAPTER 1 INTRODUCTION TO LARGE LANGUAGE MODELS IN FINANCE

of a sequence of words. They are trained on massive datasets of text and code, learning the intricate patterns and relationships within human language. Unlike earlier rule-based systems or simpler statistical models, LLMs use deep learning techniques, specifically neural networks with many layers (hence "deep"), to capture the nuances of language in a way that allows them to perform a wide range of tasks. Imagine an AI model trained on financial reports, market data, and regulatory texts—this model could generate insights, answer complex questions, or even draft financial summaries. By capturing the intricacies of language patterns, LLMs open doors to applications that streamline analysis, enhance decision-making, and mitigate risks across the finance sector.

Key Characteristics of LLMs:

- **Scale:** The "large" in LLM refers to both the size of the training dataset (often trillions of words) and the number of parameters in the model (the weights and biases that determine its behavior). This scale is crucial for achieving high performance.

- **Generative Capabilities:** LLMs can generate new text that is often indistinguishable from human-written text. This includes tasks like writing articles, summarizing documents, translating languages, and generating different kinds of creative content.

- **Contextual Understanding:** LLMs can understand the context of a given input and generate relevant and coherent responses. This is achieved through mechanisms like attention, which allows the model to focus on the most relevant parts of the input when generating output.

- **Few-Shot Learning:** Some advanced LLMs can perform well on new tasks with only a few examples (or even zero examples in some cases, known as zero-shot learning). This adaptability is a significant advantage.

Underlying Mechanisms

The dominant architecture behind modern LLMs is the *transformer*. Introduced in the Google paper "Attention Is All You Need," transformers rely heavily on the *attention mechanism*. This mechanism allows the model to weigh the importance of different words in a sequence when processing it. Unlike recurrent neural networks (RNNs), which process words sequentially, transformers can process all words in parallel, significantly speeding up training.

Here's a simplified breakdown as shown in Figure 1-1:

1. **Tokenization:** Input text is broken down into smaller units called tokens (words or sub-word units).

2. **Embeddings:** Each token is converted into a numerical vector (embedding) that represents its semantic meaning.

3. **Attention Mechanism:** The model uses attention to calculate the relationships between all tokens in the input sequence. This allows it to understand the context and dependencies between words.

4. **Feed-Forward Networks:** The attention output is passed through feed-forward neural networks to further process the information.

5. **Output Generation:** The model generates output tokens one at a time, based on the processed input and the previously generated tokens.

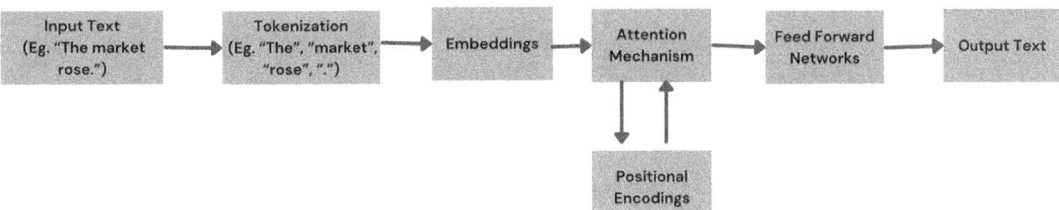

Figure 1-1. *LLM Architecture*

Figure 1-1 provides an overview of the typical LLM processing pipeline, from input text to output generation. The model tokenizes input, converts tokens into vector embeddings, applies positional encodings, and processes them through attention and feedforward layers to produce coherent output.

CHAPTER 1 INTRODUCTION TO LARGE LANGUAGE MODELS IN FINANCE

Understanding the architecture of a large language model (LLM) is essential for grasping how it processes and generates human-like text. The typical LLM pipeline involves several key components, starting from raw input text, progressing through tokenization, embeddings, and attention mechanisms, and culminating in the generation of contextually relevant output. Each stage contributes to the model's ability to interpret and produce language with nuanced understanding and fluency.

Key Advancements in LLM Technology

Several key breakthroughs have contributed to the rapid advancement of LLMs:

- **Transformer Architecture:** As mentioned above, the transformer architecture revolutionized NLP by enabling parallel processing and improving contextual understanding.

- **Increased Model Size and Data:** Training models on increasingly larger datasets and with more parameters has consistently led to performance improvements.

- **Transfer Learning:** Pre-training LLMs on massive general-purpose datasets and then fine-tuning them on smaller, task-specific datasets has proven highly effective. This allows models to learn general language patterns and then adapt them to specific tasks or domains like finance.

- **Self-Supervised Learning:** Training LLMs on unlabeled text data using self-supervised learning techniques (like predicting masked words) has enabled them to learn rich language representations without requiring expensive manual annotation.

- **Reinforcement Learning from Human Feedback (RLHF):** Fine-tuning LLMs using human feedback has improved their ability to generate more helpful, harmless, and aligned outputs.

Capabilities and Limitations

LLMs offer remarkable capabilities, including

- **Text Generation and Summarization:** Creating human-quality text and condensing large documents.

- **Language Translation:** Accurately translating between languages.

- **Question Answering:** Answering questions based on provided context.

- **Code Generation:** Writing code in various programming languages.

However, they also have limitations:

- **Lack of Real-World Understanding:** LLMs are trained on text data and don't have real-world experiences. This can lead to factual inaccuracies and nonsensical outputs in some cases.

- **Bias and Fairness:** LLMs can inherit biases present in their training data, leading to unfair or discriminatory outputs.

- **Interpretability:** Understanding why an LLM generates a particular output can be challenging. This lack of transparency can be problematic in high-stakes applications like finance.

- **Computational Cost:** Training and deploying large LLMs requires significant computational resources.

Understanding these capabilities and limitations is crucial for effectively applying LLMs in finance and mitigating potential risks. To illustrate the potential impact of LLMs on financial operations, Table 1-1 compares traditional methods with LLM-based approaches across several key financial tasks. This comparison considers factors such as accuracy, speed, and cost-effectiveness. It's important to note that the accuracy of LLMs can vary depending on the specific task, the quality of training data, and the model's architecture. Therefore, the 'Accuracy' column uses qualifiers like 'Potentially lower' or 'Can be inconsistent' to reflect this nuance. Similarly, cost-effectiveness is assessed in the long run, considering the initial investment in LLM technology vs. the ongoing costs of manual labor or traditional systems.

CHAPTER 1 INTRODUCTION TO LARGE LANGUAGE MODELS IN FINANCE

Table 1-1. Comparison of Traditional Methods vs. LLM-Based Approaches in Finance

Task	Traditional Method	LLM-Based Approach	Accuracy	Speed	Cost-Effectiveness
Document Review	Manual review by human analysts	Automated extraction and summarization by LLMs	Lower (prone to human error)	Slower	Higher
Risk Assessment	Rule-based systems, statistical models (e.g., logistic regression)	LLM-based sentiment analysis, anomaly detection	Potentially lower, less nuanced	Faster, can handle more data	Potentially lower in the long run
Compliance Reporting	Manual data collection and report generation	Automated report generation from various data sources	Lower (prone to human error)	Significantly Faster	Lower in the long run
Customer Service	Human agents, rule-based chatbots	LLM-powered chatbots, personalized responses	Can be inconsistent, limited context	Faster for routine inquiries	Potentially lower
Fraud Detection	Rule-based systems, simple anomaly detection methods	LLM-based pattern recognition, anomaly detection within complex contexts	Potentially lower for complex fraud	Faster at analyzing large datasets	Potentially lower in the long run

LLM Market Growth and Global Adoption

The rise of LLMs is reflected in the global growth of the NLP market, which has expanded dramatically as more organizations incorporate these advanced AI technologies. According to a report by MarketsandMarkets, the NLP market is projected to grow significantly, underscoring the critical role of LLMs and NLP technologies across various sectors, including finance.

Factors Driving Growth:

- **Data Availability:** The exponential growth of financial data from transactions, reports, and real-time market data has created an urgent need for tools that can process and analyze large volumes of unstructured information quickly and accurately.

- **Demand for Automation:** As digital transformation accelerates, financial institutions are increasingly seeking AI-driven solutions to automate labor-intensive tasks, streamline operations, and improve productivity.

- **Cost Efficiency:** By automating routine tasks, LLMs help organizations reduce costs associated with manual processes, making them an attractive investment for firms looking to optimize operational budgets.

Efficiency Gains in Finance: Reducing Repetitive Tasks

Efficiency is at the core of LLM adoption in finance, where the automation of repetitive tasks can lead to substantial productivity gains. AI-driven automation, including LLMs, can significantly reduce time spent on repetitive tasks in finance. This demonstrates how AI and LLMs are transforming the way financial professionals work, allowing them to shift focus from routine tasks to more strategic, value-driven activities.

Examples of Efficiency Gains:

- **Data Extraction and Document Analysis:** Tasks like extracting data from financial reports or processing compliance documents are time-consuming and error-prone. With LLMs, these processes can be automated, ensuring faster and more accurate outcomes. For instance, an LLM could review hundreds of regulatory documents, identify relevant clauses, and summarize the key information within minutes.

- **Enhanced Customer Support:** Customer service in finance often involves handling a large volume of standard inquiries. By implementing AI-powered chatbots and LLMs, institutions can automate responses to routine questions, freeing up customer service representatives to address more complex client needs. This leads to faster response times, increased customer satisfaction, and better allocation of human resources.

- **Personalized Financial Advice:** LLMs can analyze customer data and transaction history to provide tailored financial advice, helping clients make informed investment decisions. Automating these interactions can enhance customer engagement without overburdening financial advisors.

CHAPTER 1 INTRODUCTION TO LARGE LANGUAGE MODELS IN FINANCE

Strategic Benefits of Efficiency Gains

The efficiency gains from automating repetitive tasks translate into more time for employees to focus on analysis, strategic planning, and higher-value tasks. For example:

- Risk managers can dedicate more time to identifying emerging threats and trends instead of manually sifting through transaction data.

- Financial analysts can focus on interpreting data and creating actionable insights, rather than spending hours consolidating information.

- Compliance teams can shift from data collection to more proactive, strategic roles in monitoring regulatory changes and potential risks.

As a result, the productivity boost from LLM-driven automation contributes to a more agile, responsive financial institution capable of meeting client demands and navigating market changes with confidence.

Examples:

- **Automating Compliance Reporting:** Imagine a financial institution tasked with producing weekly compliance reports that outline adherence to regulatory standards. With an LLM trained on regulatory language and company policy, the process could be fully automated. The model could review transactions, flag potential issues, and generate a summary report for compliance officers, saving time and reducing human error.

- **Sentiment Analysis on Earnings Calls:** Using LLMs for sentiment analysis, financial firms can analyze the tone and context of executives' words in earnings calls, gaining insights into market sentiment. For example, an LLM trained to detect positive or negative tones in financial contexts could analyze phrases like "unprecedented growth" or "anticipated challenges," helping investors gauge market sentiment and make informed decisions.

Applications in Finance

Large language models (LLMs) are transforming the finance industry, enabling advanced applications that streamline complex processes, enhance decision-making, and improve risk management. By leveraging natural language processing (NLP), these models can understand and analyze vast amounts of financial data, transforming it into actionable insights with unprecedented speed and accuracy. This section explores key applications of LLMs in finance, illustrating how they drive operational efficiency and strategic advantage across various domains. Table 1-2 provides a concise overview of the various applications discussed, categorized by their primary function.

Table 1-2. LLM Applications in Finance—A Summary

Application Area	Specific Use Cases	Key Benefits
Automated Analysis	Document analysis and information extraction, sentiment analysis, market research and competitive intelligence, automated report generation	Increased efficiency, reduced manual effort, faster insights, improved accuracy in data extraction.
Risk Management	Credit risk assessment, fraud detection, operational risk management, compliance monitoring	Enhanced accuracy in risk assessments, faster detection of fraudulent activities, improved compliance adherence, proactive risk mitigation.
Financial Forecasting & Strategic Decisions	Market forecasting, scenario planning, algorithmic trading, personalized financial advice	More accurate forecasts, data-driven strategic decisions, automated trading strategies, personalized customer experiences.
Customer Service & Communication	AI-powered chatbots, personalized communication	Improved customer satisfaction, faster response times, 24/7 availability, personalized interactions.

Automated Analysis and Insights

LLMs excel at automating the analysis of large volumes of unstructured financial data, such as reports, earnings call transcripts, and news articles. By using their advanced contextual understanding, LLMs can extract critical information and present actionable insights with exceptional speed and precision.

- **Document Analysis and Information Extraction**

    LLMs can process complex documents, such as regulatory filings, contracts, and financial reports, to extract key metrics, identify risks, and summarize critical points.

    Example: An LLM trained on earnings reports can identify and summarize key financial metrics such as revenue growth, operating margins, and forward guidance, enabling analysts to focus on strategic interpretations rather than manual data extraction.

- **Sentiment Analysis**

    By analyzing news articles, social media, and market commentary, LLMs can assess market sentiment to uncover emerging trends and shifts in public perception.

    Example: An LLM analyzing social media posts about a company can identify whether public sentiment is trending positively or negatively, providing traders with actionable insights for real-time decision-making.

- **Market Research and Competitive Intelligence**

    LLMs can distill information from industry reports, competitor filings, and market data to offer insights into competitive dynamics and market opportunities.

    Example: An LLM scanning financial news can summarize trends in a specific sector, helping businesses identify growth opportunities and potential threats.

Risk Management

Risk management is a cornerstone of financial operations, and LLMs provide tools to proactively identify, assess, and mitigate risks with unprecedented efficiency. As shown in Figure 1-2, the following are the risks that are crucial in the finance domain:

- **Credit Risk Assessment**

    By processing data from financial statements, credit reports, and macroeconomic indicators, LLMs enhance the accuracy of creditworthiness assessments.

    Example: A bank could use an LLM to identify early warning signals in a client's financial health by analyzing public statements and credit histories, enabling timely intervention.

- **Fraud Detection**

    LLMs are adept at detecting fraudulent activities by identifying anomalies in transaction data and textual communications.

CHAPTER 1   INTRODUCTION TO LARGE LANGUAGE MODELS IN FINANCE

Example: An LLM trained on historical fraud data could detect subtle patterns indicative of insider trading or money laundering, flagging them for further review.

- **Operational Risk Management**

    By analyzing internal incident reports and policy documents, LLMs can highlight operational risks and recommend mitigation strategies.

    Example: An LLM could scan risk management logs to identify recurring operational issues and suggest preventative measures.

- **Compliance Monitoring**

    Financial institutions can use LLMs to monitor regulatory updates and ensure adherence to compliance requirements.

    Example: An LLM can identify changes in global financial regulations, summarize their implications, and flag areas where current policies may need adjustments.

Figure 1-2. LLM Applications in Risk Management

Figure 1-2 illustrates the role of LLMs in three key risk domains: credit risk (e.g., default prediction and credit scoring), fraud detection (e.g., anomaly detection and pattern recognition), and operational risk (e.g., incident analysis and process improvement).

Financial Forecasting and Predictive Analysis

By combining historical data with contextual insights, LLMs enable more accurate forecasting and predictive analysis, supporting strategic financial decisions.

- **Market Forecasting**

    LLMs analyze historical trends, economic indicators, and sentiment data to produce comprehensive market forecasts.

    Example: An LLM could predict market trends by integrating news sentiment, historical price movements, and global economic signals.

- **Scenario Planning**

    LLMs simulate various market conditions, enabling financial institutions to develop contingency plans for different scenarios.

    Example: A portfolio manager could use an LLM to model the impact of economic downturns on investment portfolios and adjust strategies accordingly.

- **Algorithmic Trading**

    Advanced LLMs can develop sophisticated trading algorithms by identifying market inefficiencies and responding in real time.

    Example: An LLM might detect arbitrage opportunities across multiple markets and execute trades instantaneously.

Customer Service and Personal Finance Assistance

LLMs redefine customer engagement in finance by delivering personalized, efficient, and accurate service.

- **AI-Powered Chatbots**

    LLMs enable chatbots to handle complex queries and provide tailored financial advice.

    Example: A chatbot powered by LLMs can guide a customer through the mortgage application process, providing personalized recommendations based on the customer's financial data.

- **Personalized Communication**

    Financial institutions use LLMs to craft personalized messages and reports, enhancing customer satisfaction.

    Example: An LLM could generate custom investment summaries for clients, highlighting performance metrics aligned with their goals.

Hybrid AI Approaches in Finance

- While large language models (LLMs) have shown remarkable capabilities in natural language understanding and generation, most financial institutions do not rely on pure LLM-based systems in isolation. Instead, they adopt hybrid AI architectures that integrate LLMs with symbolic reasoning, domain-specific rules, and traditional quantitative models. This layered approach enhances both accuracy and interpretability, aligning advanced AI capabilities with the stringent demands of financial applications.

- **Knowledge Graphs + LLMs**

    Financial reasoning often requires a structured understanding of entities, relationships, and time-bound transactions. By combining LLMs with knowledge graphs, organizations can ground model outputs in a structured context—enabling more accurate entity resolution, document linking, and contextual inference across unstructured and structured data sources. This synergy is particularly useful in areas like ESG analysis, KYC, and financial fraud detection.

- **Symbolic AI + LLMs**

  For tasks that demand logic-driven decisions, such as regulatory compliance or rule-based risk assessments, symbolic AI offers transparency and rule enforcement. When integrated with LLMs, it provides a framework where LLMs handle ambiguous or linguistic inputs, while symbolic systems enforce deterministic business rules—ensuring that AI-enhanced decisions remain auditable and explainable.

- **LLMs + Traditional Risk Models**

  Financial institutions have long relied on statistical tools like Monte Carlo simulations and Value at Risk (VaR) models. Rather than replacing these methods, LLMs can enhance them by generating scenario narratives, explaining model assumptions, or enriching input data with sentiment or macroeconomic indicators. This results in a more holistic and informed risk forecasting pipeline.

  By using these hybrid approaches, financial firms benefit from the adaptability and language understanding of LLMs while maintaining the predictability, control, and compliance required by regulatory and operational standards.

Fraud Detection and Regulatory Compliance

LLMs help ensure compliance and detect fraudulent activity with greater accuracy than traditional methods.

- **Anomaly Detection**

  By analyzing transaction data and communication logs, LLMs identify patterns indicative of fraud.

  Example: An LLM could flag unusual transaction patterns in real time, enabling swift action to mitigate risks.

- **Regulatory Monitoring**

  LLMs streamline compliance by summarizing regulatory updates and assessing the institution's adherence to them.

Example: An LLM can monitor financial communications for language that may breach regulatory standards, helping institutions avoid penalties.

Benefits of Using LLMs in Finance

Large language models (LLMs) are redefining how financial institutions operate by enhancing efficiency, accuracy, and analytical capabilities. These advanced AI systems provide transformative solutions across various financial applications, enabling organizations to streamline workflows, deliver superior client experiences, and manage risks proactively. The key benefits of LLMs in finance are discussed in the following sections.

Enhanced Efficiency and Productivity

Efficiency is a cornerstone of financial operations, and LLMs excel in automating repetitive, time-consuming tasks, allowing finance professionals to focus on strategic decision-making.

Key Advantages:

- **Automation of Routine Tasks:**
  - LLMs handle tasks such as data extraction, report generation, and customer queries with ease, significantly reducing manual effort.
  - **Example:** A regulatory compliance team uses an LLM to extract key details from lengthy documents, producing actionable summaries in minutes instead of hours.

- **Faster Processing of Large Datasets:**
  - LLMs process vast amounts of structured and unstructured data at high speed, enabling real-time analysis.
  - **Example:** An investment bank employs an LLM to analyze years of transaction data in seconds, identifying historical trends and patterns.

- **24/7 Availability for Customer Support:**

  - LLM-powered chatbots provide instant, accurate responses to routine queries, improving customer satisfaction while reducing response times.

  - **Example:** A wealth management firm deploys an LLM chatbot to handle inquiries about account balances and investment options, reducing the load on human agents.

Improved Accuracy and Reduced Human Error

In finance, even minor errors can lead to substantial financial and reputational losses. LLMs mitigate this risk by automating processes with precision and consistency.

Key Advantages:

- **Reduced Errors in Data Processing:**

  - Automating tasks like data entry and reconciliation eliminates human errors that could impact financial outcomes.

  - **Example:** An accounting team uses an LLM to match transaction records with invoices, ensuring accuracy in financial reporting.

- **Consistent and Objective Analysis:**

  - LLMs apply uniform logic to data analysis, avoiding biases and inconsistencies inherent in human judgment.

  - **Example:** A trading platform uses an LLM to detect anomalies in transaction data, such as unusual volumes or outliers, maintaining high accuracy levels.

- **Enhanced Risk Assessment:**

  - LLMs analyze large datasets to identify risks with greater precision, leading to informed decision-making.

  - **Example:** A credit risk model powered by an LLM predicts borrower default probabilities with superior accuracy by analyzing historical repayment patterns and economic conditions.

Advanced Analytical Capabilities

LLMs unlock new possibilities for extracting insights and understanding complex data, providing finance professionals with unparalleled analytical tools.

Key Advantages:

- **Sentiment Analysis:**

  - LLMs analyze news articles, social media, and reports to assess market sentiment and identify emerging trends.

  - **Example:** An investment firm uses an LLM to detect a surge in positive sentiment about renewable energy stocks, driving a profitable investment decision.

- **Contextual Understanding of Financial Documents:**

  - LLMs comprehend the nuanced language in contracts, regulatory filings, and reports, making complex information accessible.

  - **Example:** A compliance team uses an LLM to identify potential risks in loan agreements by analyzing legal language and highlighting deviations from standard terms.

- **Anomaly Detection and Fraud Prevention:**

  - LLMs analyze transaction data to detect irregular patterns or behaviors that might indicate fraud.

  - **Example:** A payment processor employs an LLM to flag transactions with unusual patterns, reducing fraud losses.

- **Scenario Planning and Forecasting:**

  - LLMs simulate multiple market scenarios to evaluate potential outcomes and guide strategic decisions.

  - **Example:** A central bank uses an LLM to model the economic impact of interest rate changes under varying market conditions.

Real-Time Risk Management

LLMs excel in processing real-time data to identify and respond to risks as they arise, a critical capability in today's volatile financial markets.

Key Advantages:

- **Monitoring Emerging Risks:**

  - LLMs analyze real-time data streams, such as news, social media, and economic indicators, to detect potential threats.

  - **Example:** A bank's risk management system uses an LLM to flag early signs of an economic downturn based on global inflation news.

- **Proactive Alerts:**

  - LLMs provide timely notifications to enable swift action on identified risks.

  - **Example:** A hedge fund receives alerts from an LLM monitoring regulatory changes, allowing them to adjust compliance strategies ahead of competitors.

Enhanced Client Experience and Personalization

LLMs enable financial institutions to offer tailored services, fostering trust and improving customer loyalty.

Key Advantages:

- **Personalized Product Recommendations:**

  - LLMs analyze client data to provide customized financial advice and product suggestions.

  - **Example:** A wealth manager uses an LLM to design a portfolio aligned with a client's financial goals and risk tolerance.

- **Proactive Client Engagement:**

  - LLMs send timely alerts and updates based on individual client preferences and transaction history.

  - **Example:** An LLM notifies a client about a favorable refinancing opportunity for their mortgage, improving satisfaction and engagement.

- **Efficient Client Support:**

  - LLMs power virtual assistants that address client concerns instantly, reducing wait times and enhancing satisfaction.

  - **Example:** A retail bank's chatbot resolves 80% of customer queries autonomously, leaving human agents to handle complex issues.

Improved Regulatory Compliance and Reporting

Compliance is a non-negotiable aspect of finance, and LLMs streamline processes to meet regulatory requirements effectively.

Key Advantages:

- **Automated Compliance Monitoring:**

  - LLMs review transactions, communications, and documents for potential compliance risks, flagging issues early.

  - **Example:** A financial institution employs an LLM to monitor emails for signs of insider trading, generating detailed reports for compliance officers.

- **Efficient Report Generation:**

  - LLMs automate the preparation of compliance reports, ensuring accuracy and timeliness.

  - **Example:** An LLM compiles transaction logs into a formatted regulatory report, saving compliance teams hours of manual effort.

CHAPTER 1   INTRODUCTION TO LARGE LANGUAGE MODELS IN FINANCE

Challenges in Financial Applications

Large language models (LLMs) offer immense potential to transform the financial industry, but their deployment comes with unique and complex challenges. From addressing data privacy concerns to ensuring compliance and maintaining operational efficiency, these challenges require careful planning and mitigation strategies. This section explores the primary obstacles in using LLMs in financial applications, emphasizing the impact on model deployment, regulatory compliance, and ethical considerations.

Data Sensitivity and Privacy

Financial institutions deal with highly sensitive data, including personal client information and confidential transaction records. Ensuring the privacy and security of this data is paramount, particularly given the regulatory requirements governing financial data.

Key Challenges:

- **Protecting Sensitive Information:**
  - LLMs require extensive datasets for training, which can include personally identifiable information (PII) and financial transaction details.
  - **Example:** A bank uses an LLM to analyze customer transactions for personalized recommendations but must anonymize the data to comply with GDPR.

- **Data Security Risks:**
  - Breaches during training, storage, or deployment can expose sensitive data. Robust encryption and access controls are essential.
  - **Example:** An unsecured data pipeline could result in unauthorized access to transactional data used in LLM training.

- **Compliance with Privacy Regulations:**
  - Regulations like GDPR, CCPA, and PCI DSS impose strict requirements on how data is collected, stored, and processed.

- **Example:** A financial institution deploying an LLM must ensure all data used adheres to regional privacy laws, adding complexity to cross-border deployments.

Mitigations:

- **Data Anonymization:** Implement techniques like masking, tokenization, and differential privacy to ensure that sensitive information is not directly accessible.
  - **Example:** Masking customer identifiers while retaining the transaction details for training fraud detection models.
- **Encryption:** Use advanced encryption (e.g., AES-256) for data at rest and in transit to prevent unauthorized access.
- **Access Controls:** Enforce strict role-based access and multi-factor authentication for systems handling sensitive data.
- **Regular Audits:** Conduct regular security audits to ensure compliance with privacy regulations and identify potential vulnerabilities.

Model Interpretability and Transparency

The "black-box" nature of LLMs poses a significant challenge in the finance industry, where decisions must be explainable and justifiable to regulators, stakeholders, and clients.

Key Challenges:

- **Lack of Transparency:**
  - Many LLMs provide outputs without clear reasoning, making it difficult to explain decisions, especially in high-stakes contexts like loan approvals or risk assessments.
  - **Example:** A client denied a loan based on an LLM's output may demand an explanation, which the model may not be able to provide.

- **Regulatory Requirements for Explainability:**
  - Financial regulations often mandate clear, auditable decision-making processes.
  - **Example:** A compliance team struggles to justify the recommendations of an LLM used for anti-money laundering (AML) analysis.
- **Trust and Accountability:**
  - Stakeholders may be reluctant to rely on models that lack interpretability, impacting adoption and trust.
  - **Example:** A risk manager hesitates to deploy an LLM for credit scoring due to its opaque decision-making process.

Mitigations:

- **Explainable AI (XAI):** Incorporate XAI techniques to provide understandable explanations for model predictions.
  - **Example:** Use SHAP (SHapley Additive exPlanations) values to explain why an LLM flagged a transaction as fraudulent.
- **Model Simplification:** Use simpler, interpretable models for decisions requiring high accountability, supplemented by LLM insights for context.
- **Audit Logs:** Maintain detailed logs of model predictions and inputs to enable post-hoc analysis and regulatory audits.
- **Stakeholder Engagement:** Collaborate with compliance officers and regulators to develop guidelines for LLM usage.

Handling Bias and Ensuring Fairness

Bias in financial data can result in discriminatory outcomes, making it critical to ensure fairness in LLM outputs.

Key Challenges:

- **Inheriting Bias from Training Data:**
    - Historical data often contains biases that LLMs may inadvertently perpetuate.
    - **Example:** An LLM trained on biased lending data could unfairly favor certain demographics over others.

- **Fairness in Decision-Making:**
    - Ensuring equitable outcomes in areas like credit scoring, loan approvals, and hiring processes is crucial.
    - **Example:** A model used for evaluating loan applications must be audited to ensure it does not unfairly reject minority applicants.

- **Regulatory and Reputational Risks:**
    - Biased outputs can lead to regulatory backlash, legal action, and damage to brand reputation.
    - **Example:** A bank faces public criticism after an LLM-powered chatbot gives inconsistent responses based on demographic factors.

A critical step in this process is the use of bias evaluation methods and fairness metrics. These tools help detect and quantify disparities in model outcomes across different demographic groups or financial segments. Common fairness metrics include

- **Equalized Odds:** Ensures that true positive and false positive rates are similar across groups.
- **Statistical Parity:** Checks whether the likelihood of favorable outcomes is the same for different groups, regardless of actual labels.
- **Demographic Parity and Predictive Parity:** Evaluate how predictions align with group distributions.

In addition to fairness metrics, model explainability tools such as SHAP (SHapley Additive exPlanations) are critical for interpreting LLM behavior. SHAP assigns each input feature a contribution score for a specific prediction, based on cooperative game theory. These scores reflect how much each feature pushed the prediction higher or lower. In financial use cases, SHAP can reveal, for example, whether a customer's income, employment history, or credit score had disproportionate influence on a loan decision. This not only helps identify potential sources of bias but also supports regulatory compliance by offering traceable, auditable justifications for model outputs.

Other explainability tools that may complement SHAP include

- **LIME (Local Interpretable Model-Agnostic Explanations)**: Generates locally linear surrogate models to approximate the behavior of complex models around a specific prediction.

- **Integrated Gradients** (for deep learning models): Measures feature attributions by integrating gradients of the prediction output relative to inputs.

- **Attention Visualizations** (especially in transformer-based LLMs): Reveal which parts of the input text the model focused on, aiding interpretation in tasks like document classification or fraud alert explanations.

Mitigations:

- **Bias Detection:** Implement pre- and post-processing techniques to identify and quantify biases in training datasets and model outputs.
  - **Example:** Analyze demographic patterns in loan approvals to identify potential biases.

- **Fairness Constraints:** Use fairness-aware algorithms during training to ensure equitable outcomes.

- **Representative Data:** Collect diverse and representative datasets to minimize bias in training.

- **Continuous Monitoring:** Regularly evaluate model outputs for fairness and adjust training processes as needed.

Model Drift and Data Volatility

The dynamic nature of financial markets means that data patterns frequently change, requiring LLMs to adapt continuously.

Key Challenges:

- **Evolving Market Conditions:**
  - LLMs trained on static datasets may struggle to adapt to real-time changes in financial markets.
  - **Example:** A trading firm's LLM trained on pre-pandemic data fails to predict post-pandemic market trends.

- **Regular Updates and Retraining:**
  - To remain accurate, models require frequent retraining on updated data, which can be resource-intensive.
  - **Example:** A bank must update its LLM for fraud detection as new fraud patterns emerge.

- **Monitoring and Mitigating Drift:**
  - Detecting and addressing model drift in a timely manner is critical to maintaining performance.
  - **Example:** An LLM used for credit risk analysis begins to misclassify applicants due to shifts in economic conditions.

Mitigations:

- **Regular Retraining:** Schedule frequent retraining of models using updated datasets to maintain accuracy.
  - **Example:** Retrain a trading prediction model quarterly to reflect changing market conditions.

- **Real-Time Monitoring:** Implement systems to monitor model performance metrics like accuracy and precision in real time.

- **Adaptive Models:** Use online learning or reinforcement learning techniques to allow models to adapt to new data dynamically.

- **Data Validation Pipelines:** Build pipelines that flag significant shifts in data distributions to trigger retraining or manual reviews.

Compliance with Regulatory Standards

The finance industry is one of the most regulated sectors, and LLM deployments must adhere to stringent legal and ethical standards.

Key Challenges:

- **Complex Regulatory Landscape:**
  - Financial institutions must navigate a maze of local and international regulations, which can vary widely.
  - **Example:** GDPR mandates data minimization and the right to be forgotten, adding layers of complexity to data management.

- **AI Governance Requirements:**
  - Regulations increasingly demand accountability and documentation of AI systems, including how they use data and make decisions.
  - **Example:** A financial institution must document every decision made by an LLM to satisfy audit requirements.

- **Severe Penalties for Non-Compliance:**
  - Failure to meet regulatory standards can result in fines, legal actions, and reputational damage.
  - **Example:** A firm deploying an LLM without appropriate safeguards is fined under CCPA for mishandling customer data.

Mitigations:

- **Regulatory Alignment:** Consult with legal and compliance teams to align LLM workflows with relevant regulations (e.g., GDPR, PCI DSS).

- **Documentation:** Maintain comprehensive records of data processing, model training, and decision-making workflows.

- **Opt-Out Mechanisms:** Ensure users can opt out of automated decision-making as required by regulations.
  - **Example:** Enable customers to request manual reviews of LLM-powered decisions affecting their credit scores.

- **Third-Party Compliance:** Vet third-party vendors to ensure their practices comply with financial regulations.

Managing High Computational Costs

The computational requirements of LLMs can be a significant barrier, particularly for smaller financial institutions.

Key Challenges:

- **High Resource Requirements:**
  - Training and deploying LLMs demand powerful hardware, such as GPUs and TPUs, and significant storage capacity.
  - **Example:** An LLM-powered risk assessment system incurs escalating costs as its usage scales across multiple departments.

- **Cost of Maintenance and Updates:**
  - Frequent retraining and infrastructure updates add to operational costs.
  - **Example:** A financial institution explores model compression techniques to reduce the costs of running an LLM.

- **Balancing Cost and Performance:**
  - Organizations must optimize models to ensure they deliver value without excessive expenses.
  - **Example:** A fintech startup uses smaller, task-specific LLMs to reduce computational overhead while maintaining performance.

Mitigations:

- **Model Optimization:** Use techniques like quantization, pruning, or distillation to reduce model size and computational requirements.
  - **Example:** Deploy a smaller distilled version of an LLM for customer support tasks to reduce resource usage.

- **Hybrid Models:** Combine LLMs with smaller, task-specific models for less demanding tasks.

- **Cloud Resource Management:** Use cloud-based infrastructure with auto-scaling and spot instances to minimize costs.

- **Cost-Benefit Analysis:** Regularly evaluate the ROI of deploying LLMs to ensure their benefits outweigh operational expenses.

The challenges of deploying LLMs in financial applications highlight the need for robust strategies to address issues like data privacy, bias, interpretability, and computational costs. By understanding and preparing for these obstacles, financial institutions can harness the transformative potential of LLMs while minimizing risks and ensuring compliance with regulatory and ethical standards. Addressing these challenges is not only essential for the successful implementation of LLMs but also for building trust and confidence among stakeholders in the finance industry.

Conclusion

This chapter introduced large language models (LLMs) and their transformative role in the finance industry. We explored how advancements in natural language processing have enabled these models to automate complex analyses, enhance risk management, and improve decision-making processes. LLMs bring significant operational benefits by streamlining tasks, increasing productivity, and generating real-time insights. However, they also present specific challenges, particularly around data sensitivity, model transparency, and regulatory compliance. Understanding these advantages and challenges gives a balanced view of what it takes to implement LLMs effectively in finance.

As we move to the next chapter, we will focus on building the essential hardware and software infrastructure that supports LLM deployment in a financial setting. This includes understanding key technical requirements, cost considerations, and security practices. With the foundation set in this chapter, you're ready to explore the technical aspects needed to make LLMs work effectively in finance.

CHAPTER 2

Infrastructure Setup for LLMs

This chapter provides a comprehensive guide to setting up the infrastructure necessary to support large language models (LLMs) in finance, where performance, scalability, and compliance are critical. Given the intense computational requirements of LLMs and the sensitivity of financial data, establishing the right infrastructure is essential for successful deployment. The chapter begins with an in-depth look at hardware requirements, covering the processing power, storage, and memory configurations that enable high-performance LLMs to operate efficiently. Following this, readers are introduced to the software stack, focusing on libraries, frameworks, and model management tools that streamline model development, deployment, and lifecycle management.

An important consideration for financial institutions is choosing between cloud-based and on-premises infrastructure. This chapter explores the pros and cons of each, highlighting cost, scalability, and security considerations unique to financial applications. By understanding the intricacies of cloud and on-premises setups, readers can make informed decisions based on their organization's specific needs and compliance requirements. Finally, the chapter concludes with best practices for creating a resilient, efficient infrastructure, including security, redundancy, and compliance guidelines that meet financial regulatory standards.

Hardware Requirements

The success of deploying and managing large language models (LLMs) in financial systems heavily depends on the underlying hardware infrastructure. The computational and storage demands of LLMs are immense, requiring specialized hardware

CHAPTER 2　INFRASTRUCTURE SETUP FOR LLMS

configurations to ensure optimal performance, scalability, and efficiency. This section provides a detailed exploration of the processing power, storage solutions, and memory configurations necessary to meet the requirements of LLMs in the financial sector.

Processing Power: GPUs, TPUs, and CPUs

1. **GPUs (Graphics Processing Units):**

   - **Role in LLMs:**

     - GPUs excel at parallel processing, making them ideal for the matrix multiplications and operations required during LLM training and inference.

   - **Advantages for Financial LLMs:**

     - High throughput for processing large datasets quickly.

     - Efficient for complex financial applications like fraud detection and risk modeling.

   - **Popular Options:**

     - NVIDIA A100, V100, and H100 GPUs are industry-leading choices for AI workloads.

   - **Example:** A trading platform uses a cluster of NVIDIA A100 GPUs to train an LLM for real-time market sentiment analysis.

2. **TPUs (Tensor Processing Units):**

   - **Role in LLMs:**

     - Developed by Google, TPUs are specialized processors optimized for deep learning tasks, including training large-scale LLMs.

   - **Advantages for Financial LLMs:**

     - Cost-effective for high-volume computations.

     - Integrated seamlessly with Google Cloud for scalable AI solutions.

- **Use Case:**
  - Training LLMs on TPUs for natural language processing tasks, such as analyzing financial reports or regulatory filings.
- **Example:** A compliance team leverages Google Cloud TPUs to train an LLM for document summarization, reducing processing time by 50%.

3. **CPUs (Central Processing Units):**
   - **Role in LLMs:**
     - CPUs provide general-purpose computing capabilities, suitable for preprocessing tasks and lightweight inference.
   - **Advantages for Financial LLMs:**
     - Ideal for handling smaller tasks such as text tokenization and model deployment where parallelism is not critical.
   - **Use Case:**
     - CPUs can support inference for low-latency applications, such as chatbots in customer support.
   - **Example:** A banking chatbot uses Intel Xeon processors to handle customer queries at scale.

Storage Solutions

1. **Solid-State Drives (SSDs):**
   - **Advantages for LLMs:**
     - High-speed data access with low latency, crucial for loading large datasets and model weights.
     - Ideal for both training and inference stages.
   - **Use Case:**
     - Storing training datasets and pre-trained model weights.
   - **Example:** An LLM training pipeline for predicting credit risk utilizes SSDs to ensure quick access to transactional data.

CHAPTER 2   INFRASTRUCTURE SETUP FOR LLMS

2. **Hard Disk Drives (HDDs):**

    - **Advantages for LLMs:**

        - Cost-effective storage option for archiving large volumes of historical data.

        - Suitable for datasets that are infrequently accessed.

    - **Use Case:**

        - Archiving historical transaction data for compliance or future model retraining.

    - **Example:** A bank uses HDDs to store historical customer interaction logs, which are accessed periodically for model updates.

3. **Network Attached Storage (NAS) and Object Storage:**

    - **Advantages for LLMs:**

        - Scalable and shared storage solutions for distributed training.

        - Supports collaboration across multiple systems or teams.

    - **Use Case:**

        - Collaborative training pipelines for LLMs across geographically distributed data centers.

    - **Example:** A multinational financial institution uses NAS for centralized access to regulatory datasets across global offices.

Memory Configurations

1. **High-RAM Systems:**

    - **Role in LLMs:**

        - High-memory configurations are essential for handling large data batches during training and accommodating the substantial size of model parameters.

- **Advantages for Financial LLMs:**
  - Reduces training time by enabling larger batch sizes.
  - Ensures smooth execution of complex tasks, such as multi-modal learning combining text and numerical data.
- **Example:** A risk modeling LLM requires 1TB of RAM to process real-time market data for training.

2. **Distributed Memory Systems:**
   - **Role in LLMs:**
     - Distributes memory across multiple nodes to handle workloads that exceed the capacity of single machines.
   - **Advantages for Financial LLMs:**
     - Enables the training of large models without running out of memory.
     - Facilitates distributed training for faster results.
   - **Use Case:**
     - Training LLMs on transaction records from multiple financial institutions.
   - **Example:** A federated learning system for anti-money laundering (AML) uses distributed memory to train collaboratively across partner organizations.

Optimizing Hardware for Financial LLMs

Building high-performance LLM systems in finance requires more than raw compute—it demands a well-balanced infrastructure optimized for speed, scalability, and cost-efficiency. Below are key strategies and examples for optimizing hardware deployments to support training and inference at scale.

CHAPTER 2   INFRASTRUCTURE SETUP FOR LLMS

1. **Hardware Acceleration**

   Using **GPUs** or **TPUs** over CPUs for compute-heavy tasks like training, fine-tuning, and inference. These accelerators dramatically reduce training time and energy consumption.

   - **Example:** An LLM built for fraud detection is trained on NVIDIA H100 GPUs, reducing end-to-end training time by over 40% compared to a CPU-based setup. Google Cloud's TPU v5p can also be used for transformer-based models with massive parameter counts.

2. **Hybrid Storage Solutions**

   Balance performance and cost by combining storage types:

   - **SSDs (Solid-State Drives):** For low-latency access to frequently used data such as embeddings, current market feeds, and real-time transaction logs.

   - **HDDs (Hard Disk Drives):** For long-term archival of regulatory documents, historical data, or offline audit trails.

   - **Example:** SSDs store high-frequency trading data for fast read/write operations, while HDDs retain archived transaction logs for monthly compliance reviews.

3. **Clustered Compute Resources**

   Distribute workloads across clusters of compute and memory nodes to support parallel training, multi-node fine-tuning, or large-scale inference workloads.

   - **Example:** A clustered infrastructure of A100 GPUs interconnected via NVLink and InfiniBand enables distributed training of an LLM on global financial markets, reducing epoch time and improving throughput.

4. **High-Bandwidth Memory and Interconnects**

   Speed up data movement between compute units using

   - **NVLink/NVSwitch:** For faster GPU-to-GPU communication in multi-GPU setups.

- **InfiniBand:** For low-latency, high-throughput interconnects in multi-node clusters.

- **HBM (High Bandwidth Memory):** Integrated in modern GPUs to support larger batch sizes and deeper models.

These technologies reduce bottlenecks and enable efficient scaling of LLMs across distributed infrastructure.

5. **NVMe-Accelerated Pipelines**

    Incorporate **NVMe storage** for high-speed caching and data streaming to support training workflows with minimal I/O delay.

    - **Example:** NVMe drives are used to cache real-time financial news and time-series data during LLM pre-training for economic scenario modeling.

With the right combination of compute acceleration, storage tiering, and interconnect technologies, financial institutions can build scalable LLM infrastructure that meets the demands of modern AI—ensuring faster innovation, lower latency, and enterprise-grade reliability.

Software Stack

The success of deploying large language models (LLMs) in financial applications depends not only on hardware but also on the software stack. The right combination of libraries, frameworks, and management tools streamlines the LLM lifecycle, from development and training to deployment and monitoring. This section explores the essential components of the software stack required to manage LLMs effectively in financial contexts, highlighting their roles, benefits, and applications.

Libraries and Frameworks

Libraries and frameworks provide the foundational tools for developing, training, and fine-tuning LLMs. Their modularity and versatility make them indispensable for financial applications requiring high precision, scalability, and compliance.

CHAPTER 2   INFRASTRUCTURE SETUP FOR LLMS

1. **PyTorch**
   - **Description:**
     - A widely used open-source deep learning framework known for its flexibility and dynamic computation graphs.
   - **Features:**
     - Supports fine-tuning of pre-trained LLMs like BERT, GPT, and T5.
     - Efficiently handles large-scale datasets and complex financial models.
   - **Example in Finance:**
     - A bank fine-tunes a GPT-based model using PyTorch to create a customer support chatbot capable of understanding financial queries.

2. **TensorFlow**
   - **Description:**
     - Another leading deep learning framework, known for its robustness and scalability.
   - **Features:**
     - Provides high-performance APIs for training and deploying LLMs across multiple platforms.
     - TensorFlow Extended (TFX) simplifies production pipelines.
   - **Example in Finance:**
     - A trading platform uses TensorFlow to train an LLM for real-time market sentiment analysis, leveraging TensorFlow Serving for deployment.

3. **Hugging Face Transformers Library**
   - **Description:**
     - A comprehensive library offering pre-trained LLMs and tools for text-based applications.
   - **Features:**
     - Easy integration with PyTorch and TensorFlow.
     - Pre-trained models like BERT, GPT-3, and T5 are readily available.
   - **Example in Finance:**
     - A compliance team uses Hugging Face models to analyze regulatory documents and flag potential risks.

Model Management Tools

Efficient management of LLMs is critical for financial applications that demand traceability, versioning, and performance monitoring. Model management tools streamline these processes, enabling teams to focus on innovation.

1. **MLflow**
   - **Description:**
     - An open-source platform for managing the entire ML lifecycle, including experiment tracking, model packaging, and deployment.
   - **Features:**
     - Tracks hyperparameters, metrics, and model versions.
     - Facilitates seamless deployment to multiple environments, such as cloud platforms or on-premises servers.
   - **Example in Finance:**
     - A financial institution uses MLflow to track and compare different versions of an LLM designed for fraud detection.

2. **Weights & Biases (W&B)**
   - **Description:**
     - A tool for tracking experiments and visualizing model performance metrics.
   - **Features:**
     - Real-time dashboards for monitoring training progress.
     - Collaboration tools for team-based model development.
   - **Example in Finance:**
     - A trading firm uses W&B to optimize hyperparameters for an LLM predicting stock market trends.

3. **TensorBoard**
   - **Description:**
     - A visualization toolkit integrated with TensorFlow for monitoring model training and performance.
   - **Features:**
     - Tracks loss curves, learning rates, and other metrics.
     - Offers plugin support for advanced visualizations.
   - **Example in Finance:**
     - A bank uses TensorBoard to monitor the fine-tuning of a sentiment analysis model for customer feedback.

4. **MLOps and CI/CD Tooling**

   To operationalize LLMs at scale in financial institutions, CI/CD pipelines and MLOps frameworks ensure that models are deployed consistently, governed robustly, and updated efficiently.

   GitHub Actions/Jenkins
   - **Purpose:** Automate ML workflows, including model retraining, validation, testing, and deployment.

- **Application in Finance:** A CI/CD pipeline triggers nightly retraining of a credit scoring model, followed by automatic validation and promotion if performance thresholds are met.

5. **TensorFlow Extended (TFX)**

    - **Purpose:** A production-grade ML pipeline framework by Google for building scalable ML systems.

    - **Components:** TensorFlow Extended (TFX) supports production-grade machine learning workflows through a set of integrated components. These include data validation and transformation, which ensure that input data is clean, consistent, and aligned with the model's requirements. Model training and evaluation follow, using standardized modules to train the model and assess its performance using defined metrics and thresholds. Finally, the push to serving environments is carried out with built-in mechanisms for monitoring, allowing teams to track model behavior post-deployment and detect drift or degradation over time. Together, these components enable scalable, auditable, and reliable ML pipelines—critical for high-stakes financial applications.

    - **Application in Finance:** A risk analytics team uses TFX to manage an end-to-end model pipeline, ensuring compliance checks are embedded before model deployment.

Deployment and Orchestration Tools

Deployment tools enable seamless scaling of LLMs in production environments, while orchestration tools ensure efficient resource utilization.

1. **Docker**

    - **Description:**

        - A containerization platform that simplifies the deployment of LLMs across different environments.

- **Features:**
  - Isolates applications, ensuring consistency across development and production.
  - Supports GPU-enabled containers for high-performance inference.
- **Example in Finance:**
  - A bank deploys an LLM-powered chatbot in Docker containers to maintain consistency across cloud and on-premises servers.

2. **Kubernetes**
   - **Description:**
     - An orchestration platform for managing containerized applications at scale.
   - **Features:**
     - Automates scaling, deployment, and management of LLM services.
     - Supports load balancing for high-traffic applications.
   - **Example in Finance:**
     - A trading platform uses Kubernetes to deploy an LLM for real-time risk assessment, ensuring high availability during market surges.

3. **TensorFlow Serving**
   - **Description:**
     - A system for serving TensorFlow models in production.
   - **Features:**
     - Offers RESTful APIs for model inference.
     - Handles versioning and model updates seamlessly.

- **Example in Finance:**
  - A compliance tool leverages TensorFlow Serving to provide real-time regulatory risk assessments.

Data Processing and Integration Tools

Before training or inference, financial datasets must be preprocessed, cleaned, and integrated seamlessly into the LLM pipeline.

1. **Apache Spark**
   - **Description:**
     - A distributed data processing framework designed for handling large-scale datasets.
   - **Features:**
     - Processes structured and unstructured financial data efficiently.
     - Integrates with ML libraries like PySpark.
   - **Example in Finance:**
     - Processing transactional data for an LLM designed to detect suspicious activities in banking.

2. **Pandas**
   - **Description:**
     - A Python library for data manipulation and analysis.
   - **Features:**
     - Provides easy-to-use data structures for cleaning and preparing financial data.
   - **Example in Finance:**
     - Preprocessing customer data for a personalized financial advisory LLM.

3. **Airflow**

    - **Description:**

        - A workflow orchestration tool for automating data pipelines.

    - **Features:**

        - Schedules and monitors data ingestion and preprocessing workflows.

    - **Example in Finance:**

        - Automating the preprocessing pipeline for a credit scoring LLM.

The software stack for managing LLMs in financial contexts must be robust, scalable, and tailored to specific use cases. Libraries like PyTorch and TensorFlow form the foundation for model development, while tools like MLflow and Docker streamline deployment and lifecycle management. Data processing tools such as Apache Spark ensure that financial datasets are efficiently prepared for training and inference. By leveraging the right combination of these components, organizations can build efficient, compliant, and high-performing LLM workflows, meeting the unique demands of the financial sector.

Cloud vs. On-Premises Solutions

Selecting the right infrastructure for deploying and managing large language models (LLMs) is critical in financial applications where performance, scalability, security, and compliance play pivotal roles. Organizations must choose between cloud-based and on-premises infrastructure—or a hybrid model—based on their unique needs. This section provides a comparative analysis of cloud and on-premises solutions, focusing on cost, scalability, and security, with practical examples to guide financial institutions in making informed decisions.

Cloud Infrastructure

Cloud-based solutions provide on-demand access to computing resources through services like Amazon Web Services (AWS), Google Cloud Platform (GCP), and Microsoft Azure. These platforms offer robust scalability and a range of managed services tailored to LLM operations.

Advantages of Cloud Infrastructure

1. **Cost-Effectiveness**

    - **Pay-As-You-Go Model:** Organizations pay only for the resources they consume, eliminating the need for large upfront investments.
    - **Reduced Maintenance Costs:** The cloud provider manages hardware, updates, and maintenance, reducing operational overhead.
    - **Example in Finance:**
        - A bank uses GCP for training an LLM-powered risk assessment tool, avoiding the costs of procuring and maintaining high-performance hardware.

2. **Scalability**

    - **Elastic Scalability:** Cloud platforms allow seamless scaling of resources up or down based on workload demands.
    - **Handling Spikes:** Ideal for handling unpredictable spikes in usage, such as high-frequency trading during market turbulence.
    - **Example in Finance:**
        - A trading platform scales its cloud infrastructure during major economic announcements to process real-time sentiment analysis.

3. **Speed and Agility**

    - **Rapid Deployment:** Cloud platforms enable quick provisioning of resources, accelerating LLM development and deployment timelines.
    - **Access to Advanced Features:** Providers offer pre-configured AI services like Google's Vertex AI and AWS SageMaker, streamlining workflows.
    - **Example in Finance:**
        - A financial institution deploys an LLM-based compliance chatbot within days using pre-built services on Azure.

Challenges of Cloud Infrastructure

1. **Security and Compliance**

   - Shared responsibility for security requires meticulous management of access controls and encryption.
   - Data residency laws may restrict where financial data can be stored or processed.
   - **Example in Finance:**
     - A European bank must ensure that customer data used for LLM training complies with GDPR, which may require hosting data in EU-based servers.

2. **Cost Management**

   - Over-reliance on the cloud can lead to escalating costs, especially for continuous, large-scale LLM training.
   - **Example in Finance:**
     - A bank incurs unexpected expenses due to prolonged training sessions on large datasets.

On-Premises Infrastructure

On-premises solutions involve hosting and managing infrastructure within an organization's own data centers. This model offers greater control over resources and data but comes with significant capital and operational requirements.

   Advantages of On-Premises Infrastructure

   1. **Security and Control**

      - **Enhanced Security:** Organizations have full control over data and hardware, reducing exposure to third-party risks.
      - **Compliance Simplification:** Easier to meet stringent regulatory requirements by hosting data locally.

- **Example in Finance:**
  - A central bank uses on-premises servers to train LLMs on confidential economic datasets, ensuring complete isolation from external networks.

2. **Cost Predictability**

One of the key considerations when deploying large language models (LLMs) in finance is long-term cost efficiency. On-premises infrastructure offers the advantage of fixed costs; after the initial capital expenditure on hardware and setup, operational expenses remain predictable and often lower than ongoing cloud usage for high-throughput workloads.

- **Example in Finance:**
  - A credit bureau implements on-prem infrastructure to support LLM training for credit risk modeling. Over several quarters, the organization sees reduced compute costs compared to variable cloud billing during peak training cycles.

To make informed infrastructure choices, it's essential to consider the Total Cost of Ownership (TCO). TCO models provide a structured way to compare cloud-based and on-premises setups, taking into account not just upfront and recurring costs, but also factors like data egress fees, system maintenance, compliance overhead, scalability needs, and long-term flexibility. Introducing TCO analysis into this decision-making process helps ensure that cost predictability aligns with organizational goals and regulatory requirements, making this a critical lens for financial institutions evaluating deployment strategies.

3. **Customization**
   - Organizations can design and optimize infrastructure for specific workloads.
   - **Example in Finance:**
     - A hedge fund customizes its on-premises cluster to optimize LLM inference for real-time trading strategies.

Challenges of On-Premises Infrastructure

1. **High Upfront Costs**

   - Procuring and setting up hardware involves significant capital expenditure.

   - **Example in Finance:**

     - A startup struggles to afford the initial investment required for GPU clusters.

2. **Scalability Limitations**

   - Scaling on-premises systems requires purchasing and configuring additional hardware, which can delay projects.

   - **Example in Finance:**

     - A financial institution cannot quickly accommodate a sudden increase in LLM training demand due to hardware constraints.

3. **Maintenance Overheads**

   - Managing hardware and software updates and ensuring uptime require dedicated IT teams and resources.

   - **Example in Finance:**

     - A bank experiences downtime due to delays in replacing faulty on-premises hardware components.

Choosing between cloud and on-premises infrastructure is a strategic decision that impacts cost, scalability, compliance, and operational agility, particularly in financial institutions deploying large language models (LLMs). Each option presents unique advantages and trade-offs depending on the organization's data sensitivity, regulatory obligations, and workload intensity. Table 2-1 offers a side-by-side comparison across key dimensions to help stakeholders evaluate the best fit for their LLM deployment strategy. Table 2-1 highlights distinctions in cost structure, scalability, security, compliance, setup time, and customization—providing a framework for Total Cost of Ownership (TCO) assessments and infrastructure planning.

Table 2-1. Comparative Analysis of Cloud vs. On-Premises Infrastructure for LLM Deployment in Finance

Aspect	Cloud Infrastructure	On-Premises Infrastructure
Cost	Pay-as-you-go; lower upfront costs but variable long-term expenses.	High initial investment but predictable operating costs.
Scalability	Instant scalability to handle spikes in demand.	Limited by physical hardware; slower to scale.
Security	Requires robust management of shared responsibilities.	Full control over security and data residency.
Compliance	Provider certifications required (e.g., GDPR, PCI DSS).	Easier to enforce compliance with local regulations.
Setup Time	Rapid deployment with pre-built services.	Time-intensive procurement and setup process.
Customization	Limited to provider offerings.	Fully customizable to meet specific workload and regulatory needs.

Hybrid Solutions

Many organizations adopt a hybrid approach, combining the strengths of cloud and on-premises infrastructure to meet specific needs.

- **Use Case:**
  - On-premises infrastructure is used for storing sensitive data and ensuring compliance, while the cloud is leveraged for scalable LLM training and deployment.

- **Example in Finance:**
  - A multinational bank uses on-premises servers for customer data storage and Google Cloud TPUs for training an LLM to detect global transaction anomalies.

The choice between cloud and on-premises infrastructure depends on an organization's priorities, including cost, scalability, security, and compliance requirements. While the cloud offers flexibility, scalability, and speed, on-premises infrastructure provides enhanced security, control, and cost predictability for long-term projects. Financial institutions must carefully evaluate their needs and consider hybrid solutions to balance the benefits of both approaches, ensuring robust and efficient LLM operations tailored to their unique challenges.

Best Practices for Infrastructure Setup

Establishing a robust and efficient infrastructure for deploying and managing large language models (LLMs) in financial contexts requires adherence to best practices. These practices are essential for ensuring security, redundancy, and compliance, particularly given the sensitive nature of financial data and the stringent regulatory landscape. This section outlines detailed guidelines that financial institutions can follow to optimize their infrastructure for security, reliability, and legal adherence.

Security Guidelines

Security is paramount in financial infrastructure setups to protect sensitive data and prevent unauthorized access. Best practices include

a. **Encryption**

- **Data at Rest:**
    - Encrypt all stored data using industry-standard encryption algorithms like AES-256 to ensure it remains inaccessible in the event of a breach.
    - **Example:** Encrypting transactional data used for LLM training on fraud detection.
- **Data in Transit:**
    - Use protocols such as TLS (Transport Layer Security) to secure data transmission between servers and clients.
    - **Example:** Securing API calls between an LLM-powered compliance tool and a regulatory database.

b. **Access Controls**

- **Role-Based Access Control (RBAC):**
  - Assign access permissions based on roles, ensuring that employees access only the data and systems necessary for their tasks.
  - **Example:** Restricting access to customer transaction data to authorized compliance officers only.
- **Multi-Factor Authentication (MFA):**
  - Implement MFA to add a layer of security for accessing critical systems.
  - **Example:** Requiring both a password and biometric verification for infrastructure administrators.

c. **Network Security**

- **Firewalls and Intrusion Detection Systems (IDS):**
  - Use firewalls to filter traffic and IDS to detect suspicious activity.
  - **Example:** Monitoring for unauthorized attempts to access LLM training servers.
- **Isolated Networks:**
  - Create isolated networks for sensitive workflows, such as LLM training on private financial datasets.
  - **Example:** Using a separate VLAN for LLM inference systems in a banking environment.
- **Federated Learning for Data Privacy:**
  - While traditional network isolation protects the training environment, federated learning provides an additional layer of protection by avoiding the need to centralize sensitive data. Frameworks like Google's Federated Learning and Flower allow models to be trained locally on edge devices or institutional silos, with only model updates (not raw data) shared.

- Example: A consortium of banks uses federated learning to collaboratively train an LLM on fraud detection without exposing proprietary or customer data across institutions.

This approach is particularly useful in jurisdictions with strict data residency laws, enabling privacy-preserving collaboration across institutions without compromising on model performance.

Redundancy Guidelines

Infrastructure redundancy ensures system availability and resilience, minimizing downtime and data loss in the event of failures.

a. **Data Redundancy**

- **Regular Backups:**
  - Perform scheduled backups of critical data and store them in geographically diverse locations.
  - **Example:** Backing up LLM training datasets to multiple data centers across regions.
- **Replication:**
  - Use data replication to create real-time copies of critical datasets for quick recovery.
  - **Example:** Replicating model weights and training logs to cloud storage for disaster recovery.

b. **Hardware Redundancy**

- **Failover Systems:**
  - Deploy redundant servers and storage devices to automatically take over in case of hardware failure.
  - **Example:** A failover cluster of GPUs ensures uninterrupted LLM training during hardware outages.

- **Load Balancing:**
  - Distribute workloads across multiple servers to prevent overloading and ensure consistent performance.
  - **Example:** Using a load balancer for evenly distributing inference requests to an LLM deployed in a trading application.

c. **Network Redundancy**
- **Multiple Connectivity Options:**
  - Use multiple internet service providers (ISPs) to ensure connectivity during outages.
  - **Example:** Maintaining dual ISP connections for real-time market analysis systems.
- **Redundant Paths:**
  - Create multiple network paths to avoid single points of failure.
  - **Example:** Setting up redundant paths between data centers hosting LLM infrastructure.

Compliance Guidelines

Compliance ensures that infrastructure adheres to legal and regulatory requirements, reducing the risk of penalties and reputational damage.

a. **Regulatory Framework Alignment**
- **Understanding Regulations:**
  - Identify and align infrastructure with relevant regulations like GDPR, CCPA, and PCI DSS.
  - **Example:** Ensuring that all customer data used for LLM training complies with GDPR's anonymization requirements.

- **Data Residency:**
  - Host data in regions that comply with local regulations.
  - **Example:** Storing European customer data in EU-based servers to comply with GDPR.

b. **Documentation and Audits**
- **Detailed Documentation:**
  - Maintain comprehensive documentation of data workflows, access logs, and security measures.
  - **Example:** Documenting how financial datasets are preprocessed and anonymized for LLM training.
- **Regular Audits:**
  - Conduct internal and external audits to ensure ongoing compliance with regulations.
  - **Example:** Annual compliance audits for LLM training pipelines to verify adherence to PCI DSS.

c. **Incident Response Planning**
- **Proactive Measures:**
  - Develop incident response plans to address security breaches or compliance violations.
  - **Example:** A predefined response strategy for handling data breaches involving LLM inference systems.
- **Testing:**
  - Regularly test response plans through simulations.
  - **Example:** Conducting a mock exercise to simulate a GDPR-related data breach scenario.

Best Practices for Financial-Specific Infrastructure

Financial institutions operate under unique constraints and risks that require specialized practices:

1. **Real-Time Monitoring**

    - Deploy monitoring tools to track infrastructure performance, detect anomalies, and prevent downtime.

    - **Example:** Monitoring GPU utilization during LLM training to optimize performance and prevent resource bottlenecks.

2. **Hybrid Infrastructure Solutions**

    - Combine cloud and on-premises setups to balance scalability with control over sensitive data.

    - **Example:** Hosting LLM inference systems in the cloud for scalability while storing training datasets on-premises for compliance.

3. **Data Minimization**

    - Collect and process only the data necessary for specific LLM applications to reduce exposure.

    - **Example:** Anonymizing and aggregating transaction data before using it for LLM training to detect fraud.

4. **Scalable Design**

    - Design infrastructure with scalability in mind to accommodate future growth in data and model sizes.

    - **Example:** Using Kubernetes to scale LLM services dynamically based on market activity levels.

Setting up infrastructure for LLMs in the financial sector requires a meticulous approach to security, redundancy, and compliance. By following these best practices, organizations can build resilient, secure, and compliant systems capable of handling the unique demands of financial applications. These guidelines not only safeguard sensitive data and ensure regulatory adherence but also provide the scalability and reliability necessary to support the growing role of LLMs in transforming financial operations.

Emerging Hardware Trends

As the demand for training and deploying large language models (LLMs) grows, new generations of specialized hardware are reshaping how financial institutions approach cost, scalability, and performance.

Beyond traditional GPUs and CPUs, **AI accelerators** are becoming increasingly popular for their **efficiency, speed, and energy optimization**. These purpose-built chips offer high throughput for training large models while reducing infrastructure costs and power consumption.

Key Examples

- **AWS Trainium**

    Custom-designed by Amazon for high-performance ML training on AWS. Optimized for cost-effective large-scale LLM workloads, Trainium supports deep learning frameworks like PyTorch and TensorFlow.

- **Google Cloud TPU v5p**

    Designed specifically for next-generation AI models, TPU v5p offers massive compute power and high-speed interconnects ideal for distributed LLM training in cloud environments.

- **Intel Habana Gaudi2**

    An open-ecosystem alternative to GPUs, Habana Gaudi accelerators offer competitive performance and lower TCO, especially when integrated with frameworks like PyTorch. Supported in cloud platforms like AWS EC2 DL1 instances.

These emerging solutions allow financial organizations to make **informed trade-offs between performance, cost, and flexibility**, especially when choosing between on-premise, hybrid, or cloud-based AI infrastructure.

As the LLM landscape evolves, staying current with hardware innovations is crucial for maintaining competitive AI capabilities while optimizing resource efficiency.

Conclusion

The infrastructure supporting large language models (LLMs) forms the foundation for their successful deployment in the finance industry. This chapter explored the key components necessary for building a robust and efficient infrastructure, including hardware requirements, software stack, deployment models, and best practices. By leveraging high-performance hardware such as GPUs and TPUs and utilizing essential tools like PyTorch and TensorFlow, financial institutions can address the computational demands of LLMs while ensuring scalability and efficiency. Additionally, carefully evaluating the pros and cons of cloud-based, on-premises, and hybrid setups enables organizations to meet their specific operational and compliance needs.

A well-designed infrastructure not only optimizes performance but also safeguards sensitive financial data and ensures adherence to stringent regulatory requirements. By implementing security measures, redundancy protocols, and compliance guidelines, organizations can mitigate risks and build reliable systems that support LLM-driven financial applications. This approach positions financial institutions to remain competitive by delivering innovative, secure, and efficient AI solutions in an increasingly complex and fast-paced industry.

With the infrastructure established, the next chapter transitions into the critical processes of **training and fine-tuning large language models (LLMs)** for financial applications. Training ensures that models are equipped to understand the complexities of financial language, address domain-specific tasks, and deliver accurate, context-relevant insights. This chapter will guide readers through essential topics such as data preparation, which involves collecting, cleaning, and curating high-quality datasets while maintaining privacy and reducing biases. It will also delve into advanced training strategies, including distributed training techniques and optimization methods to handle large-scale datasets efficiently.

The chapter will further explore fine-tuning pre-trained models to adapt them for specific financial tasks like fraud detection, sentiment analysis, and compliance reporting. Finally, readers will learn best practices for evaluating and validating model performance, focusing on accuracy, reliability, and relevance in real-world financial scenarios.

CHAPTER 3

Training and Fine-Tuning LLMs

Training and fine-tuning large language models (LLMs) are critical steps in adapting these models to the unique demands of the financial sector. From handling sensitive data to ensuring compliance and delivering accurate, domain-specific insights, these processes enable LLMs to perform effectively in high-stakes environments.

This chapter provides a practical guide to preparing data, training models, and fine-tuning them for financial applications such as fraud detection, credit risk assessment, and strategic forecasting. It covers best practices in data curation, distributed training, and model optimization, as well as advanced fine-tuning techniques like domain adaptation and task-specific tuning.

By the end of this chapter, you will be equipped with the knowledge and tools to build robust, scalable LLMs tailored for real-world financial use cases—addressing performance, compliance, and operational challenges with confidence.

Goals and Processes of Training and Fine-Tuning

Training and fine-tuning large language models (LLMs) are key steps in preparing them for specific applications. While training focuses on building a model from scratch or pretraining it on a general dataset, fine-tuning adapts the model to specific tasks or domains. In finance, where precision and compliance are paramount, training and fine-tuning play a crucial role in ensuring that LLMs meet high-performance standards.

This section provides an overview of the goals and processes involved in training and fine-tuning LLMs, with a focus on financial applications. It highlights the importance of high-quality data, effective methodologies, and adaptive strategies to create models that address domain-specific challenges such as fraud detection, credit scoring, and portfolio optimization.

CHAPTER 3 TRAINING AND FINE-TUNING LLMS

To effectively train and fine-tune large language models (LLMs) in financial applications, it's essential to understand the end-to-end lifecycle that governs their development and deployment. The lifecycle is not a linear pipeline but a continuous, iterative process involving five key stages: Input Data, Training, Fine-Tuning, Deployment, and Feedback Loop. Each stage plays a critical role in shaping the model's performance, adaptability, and alignment with real-world financial needs.

Figure 3-1 illustrates the core stages of the LLM lifecycle. It begins with high-quality input data, followed by initial training and domain-specific fine-tuning. After deployment, performance monitoring and user feedback feed into a continuous improvement loop, ensuring the model remains accurate, relevant, and compliant over time.

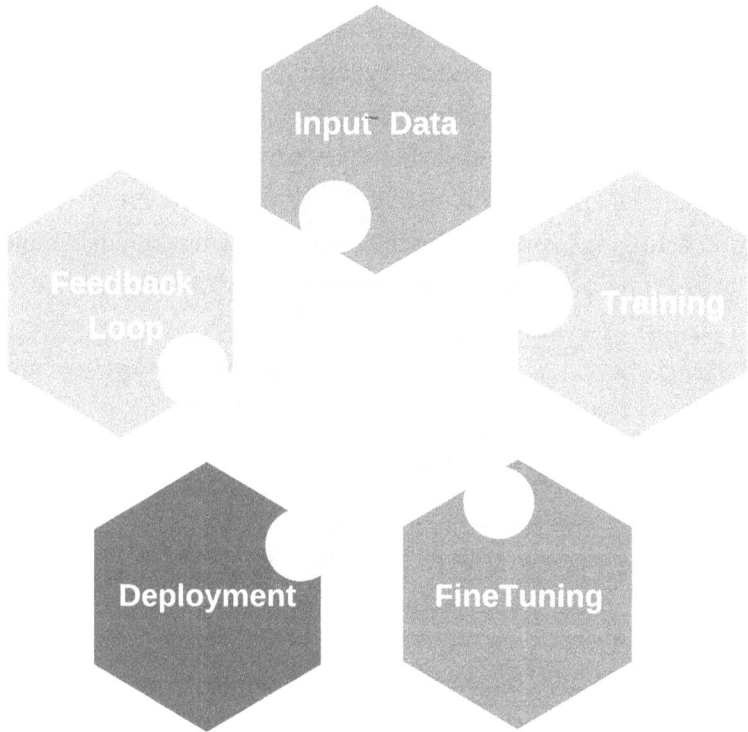

Figure 3-1. *The LLM Lifecycle*

CHAPTER 3   TRAINING AND FINE-TUNING LLMS

Training and fine-tuning large language models (LLMs) involve distinct yet interconnected processes that adapt these models to meet specific application needs. In the financial sector, the primary goal of training is to create a robust, generalized model, while fine-tuning customizes the model for domain-specific tasks. These steps are essential for ensuring that LLMs achieve precision, reliability, and compliance with industry standards.

Table 3-1 summarizes key differences in terms of dataset size, computational cost, time, and outcomes, especially relevant when adapting LLMs for financial applications.

Table 3-1. Training vs. Fine-Tuning of LLMs in Finance

Aspect	Training	Fine-Tuning
Objective	Build a generalized language model	Adapt the model to specific financial tasks or domains
Dataset Size	Very large (hundreds of GBs to TBs)	Smaller, task-specific datasets (MBs to a few GBs)
Computational Cost	Extremely high (requires distributed compute infrastructure)	Moderate (can often run on fewer GPUs or cloud instances)
Time Requirement	Weeks to months	Hours to days
Expected Outcome	Broad language understanding across domains	Task-adapted model with domain-specific accuracy
Example in Finance	Training on financial news, SEC filings, macroeconomic data	Fine-tuning for credit scoring or regulatory document summarization

As shown in Figure 3-2, LLM training creates a generalized model using large-scale, diverse data. Fine-tuning then adapts this model with domain-specific data to produce a task-adapted LLM.

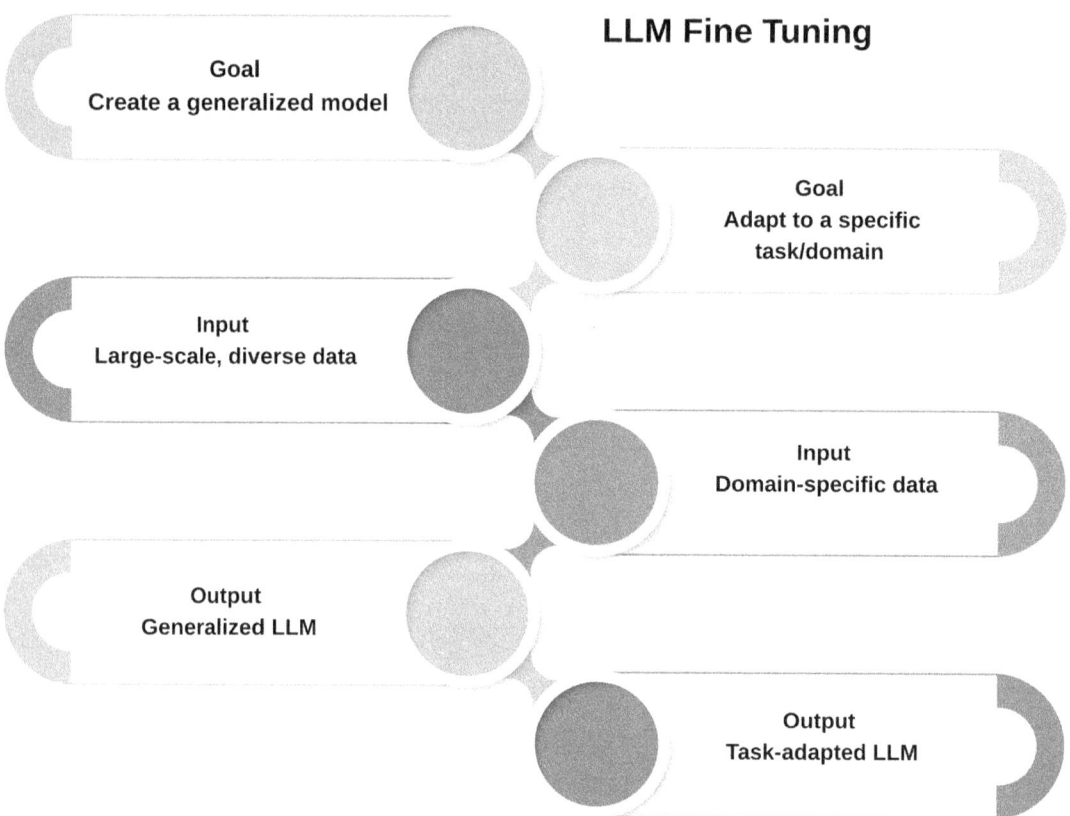

Figure 3-2. LLM Training

High-Quality Data: The Foundation of Success

In training and fine-tuning large language models (LLMs), the quality of data serves as an important factor of success. In finance, where precision, trust, and compliance are paramount, poor-quality data can compromise the reliability and interpretability of models, leading to significant financial and reputational risks. High-quality data not only ensures accuracy and performance but also helps models generalize effectively across diverse financial tasks, such as fraud detection, credit scoring, and investment optimization. Achieving this requires a meticulous approach to data collection, cleaning, and augmentation.

Data Collection: Sourcing Reliable and Diverse Financial Data

The process of data collection lays the groundwork for model training. In the financial sector, the variety and reliability of data sources are crucial to building a robust LLM. Key considerations include

- **Diverse Data Sources:**

  - **Transactional Data:** Includes anonymized payment records, credit card transactions, and loan histories. These provide granular insights for tasks like fraud detection and credit risk assessment.

  - **Financial Statements:** Publicly available reports such as balance sheets, income statements, and cash flow statements offer structured data for credit scoring and portfolio management.

  - **Market News and Sentiment Analysis Data:** Unstructured data from financial news articles, analyst reports, and social media provide context for market behavior predictions.

  - **Regulatory Filings:** Data from filings like 10-Ks and 10-Qs in the United States are essential for compliance and risk analysis models.

  - **Macroeconomic Indicators:** Data on interest rates, inflation, and GDP add valuable external context to models operating in a dynamic financial landscape.

- **Reliability of Sources:**

  Ensure data is sourced from trusted platforms and institutions, minimizing the risk of inaccuracies. Vet sources regularly to maintain data integrity.

- **Ethical Considerations:**

  Adhere to regulations such as GDPR and CCPA to ensure the ethical handling of sensitive financial data, avoiding unauthorized usage or data breaches.

Data Cleaning: Enhancing Usability and Consistency

Raw financial data often contains noise, inconsistencies, and irrelevant information that can degrade model performance. Effective cleaning processes ensure that only meaningful and high-quality data is fed into the model. Critical steps include

- **Dealing with Missing Values:**
  - Employ imputation techniques such as mean, median, or predictive modeling to fill gaps.
  - In critical financial datasets, missing values should be flagged as potential anomalies.

- **Removing Duplicates and Outliers:**
  - Detect duplicate records using unique identifiers.
  - Use statistical methods like interquartile range (IQR) or z-scores to identify and handle outliers.

- **Standardizing Formats:**
  - Normalize numerical data (e.g., currency conversions) to a common scale.
  - Standardize textual data, such as financial terms, to ensure consistency across the dataset.

- **Tokenization and Language Cleaning:**
  - Split text into tokens relevant to the financial domain (e.g., "interest rate" should not be split into separate words).
  - Remove stopwords unless they carry domain-specific significance (e.g., "rate" in "interest rate").

- **Noise Removal:**
  - Strip irrelevant data like advertisements in news articles or unstructured footnotes in financial statements that do not add value to the model.

Data Augmentation: Enriching the Dataset

Data augmentation techniques expand and balance datasets, ensuring robust training even in scenarios with limited or imbalanced data. In finance, augmentation techniques are especially valuable for handling rare events, such as fraud or market crashes, which may be underrepresented in raw datasets. Strategies include

- **Synthetic Data Generation:**
  - Create synthetic financial transaction records using generative models, such as Variational Autoencoders (VAEs) or Generative Adversarial Networks (GANs).
  - Simulate rare events like fraudulent activities or economic crises to train models for edge cases.

- **Paraphrasing and Text Expansion:**
  - Generate variations of financial text, such as rewording market analysis reports, to expand unstructured datasets.
  - Use transformers to rewrite textual data while preserving meaning.

- **Oversampling and Undersampling:**
  - Apply techniques like SMOTE (Synthetic Minority Over-sampling Technique) to balance datasets, particularly when working with rare categories, such as default loans or fraudulent transactions.

- **Data Fusion:**
  - Combine data from multiple sources (e.g., merging transactional data with sentiment analysis) to provide a richer dataset for context-aware models.

High-quality data is a crucial factor of a successful LLM training and fine-tuning in finance. From sourcing reliable data to cleaning and augmenting it, every step must be executed meticulously. These efforts ensure that the resulting model not only performs accurately but also aligns with the nuanced requirements of financial applications. By investing in the foundation of data quality, organizations can unlock the full potential of LLMs in driving financial innovation and precision.

Table 3-2 highlights the key differences between training and fine-tuning, showing how they vary in goals, data needs, resources, and outputs. It clarifies why training builds general-purpose models, while fine-tuning adapts them for specific tasks.

Table 3-2. Differences Between Training and Fine-Tuning

Aspect	Training	Fine-Tuning
Goal	Build a general-purpose model	Adapt a model to specific tasks
Data Requirements	Large, diverse datasets	Domain-specific datasets
Resource Requirements	High (time, compute power)	Moderate (dependent on model size)
Output	Generalized LLM	Task-optimized LLM

Table 3-3 summarizes common challenges in training such as overfitting, data imbalance, and high computational cost. It also provides practical solutions like regularization, data augmentation, and distributed training.

Table 3-3. Common Challenges and Solutions in Training

Challenge	Description	Solution
Overfitting	Model performs well on training data but poorly on new data	Use regularization techniques, early stopping
Data Imbalance	Uneven representation of classes in data	Data augmentation, weighted loss functions
High Computational Cost	Training requires significant resources	Use distributed training, optimize hyperparameters
Output	Generalized LLM	Task-optimized LLM

Effective Methodologies for Training and Fine-Tuning

Training and fine-tuning large language models (LLMs) are foundational to creating models that perform well across various financial applications. Structured methodologies in both processes are critical for optimizing performance, ensuring model scalability, and addressing domain-specific challenges. This section explores key techniques for both training and fine-tuning LLMs, focusing on their relevance and application in finance.

Training Methodologies

Training an LLM involves pretraining the model on large-scale, diverse datasets to help it learn generalized language patterns. This stage sets the groundwork for subsequent fine-tuning and requires careful consideration of methodologies to ensure optimal performance.

1. **Utilizing Large, Diverse Datasets**

    - **Purpose:** Pretraining on diverse datasets ensures the model develops a robust understanding of linguistic patterns, making it versatile and adaptable to various tasks.

    - **Examples of Datasets:** Wikipedia, Common Crawl, financial news archives, and open-domain corpora with structured and unstructured data.

    - **Best Practices:**
        - Include multilingual data to enhance language understanding in global financial contexts.
        - Integrate domain-relevant corpora such as SEC filings, credit reports, and financial analysis documents to align with financial terminologies.

2. **Distributed Training for Scalability**

    - **Purpose:** Large-scale LLMs require immense computational resources. Distributed training enables splitting the workload across multiple GPUs or TPUs.

- **Techniques:**
  - **Data Parallelism:** Splits data across devices while keeping the model copy identical.
  - **Model Parallelism:** Splits the model across devices to manage memory constraints.
  - **Pipeline Parallelism:** Distributes different parts of the training process across devices.
- **Cost-Aware Enhancements:**

  In financial applications, computational efficiency directly affects ROI. The following techniques are commonly used to reduce training cost and resource consumption:
  - **Gradient Checkpointing:** Saves memory by selectively storing intermediate activations and recomputing them during backpropagation.
  - **Activation Offloading:** Temporarily transfers activations to CPU memory to reduce GPU memory usage.
  - **Mixed-Precision Training (FP16/BF16):** Uses lower-precision arithmetic to reduce compute time and memory consumption, often without sacrificing accuracy.
- **Key Tools:** Frameworks like TensorFlow, PyTorch Distributed, and DeepSpeed optimize the distributed training process.

3. **Hyperparameter Optimization**
   - **Purpose:** Proper tuning of hyperparameters ensures better convergence and improved performance.
   - **Key Parameters to Optimize:**
     - **Learning Rate:** Affects the step size in updating model weights. Gradual learning rate warm-ups followed by decay schedules (e.g., cosine annealing) are common.

- **Batch Size:** Larger batch sizes improve stability in gradient updates but require more memory. Dynamic batch sizing can adapt to resource constraints.

- **Dropout Rate:** Helps prevent overfitting by randomly deactivating a fraction of neurons during training.

- **Optimization Techniques:** Grid search, random search, and advanced methods like Bayesian optimization or population-based training.

Fine-Tuning Techniques

Fine-tuning builds on the pretrained model to adapt it to specific tasks or domains, enabling it to address unique financial applications. The process involves task-specific datasets and iterative improvements for optimal alignment.

1. **Task-Specific Adaptation**

   - **Purpose:** Customize the model for applications like fraud detection, credit scoring, or portfolio optimization.

   - **Approach:**
     - Use labeled datasets relevant to the target task. For example, transactional data labeled for fraud or historical credit performance data.
     - Train the model on these datasets using supervised learning techniques to adjust weights specific to the task.

   - **Example:** Fine-tuning an LLM on a dataset of labeled suspicious transactions to enhance its fraud detection capabilities.

2. **Domain Adaptation**

   - **Purpose:** Align the model with financial terminologies, practices, and regulatory considerations.

   - **Techniques:**

- **Domain-Adaptive Pretraining (DAPT):** Continue pretraining on domain-specific corpora, such as financial filings or industry reports, before task-specific fine-tuning.

- **Feature Alignment:** Modify embeddings to capture domain-specific relationships, such as correlations between financial terms.

- **Impact:** Improves the model's ability to interpret and generate domain-specific outputs, such as financial forecasts or compliance checks.

3. **Iterative Updates for Model Refinement**

   - **Purpose:** Continuously improve the model's performance in response to new data or changing requirements.

   - **Process:**
     - Retrain the model periodically with new datasets to address shifts in financial trends or emerging risks (e.g., new types of fraud).
     - Use active learning to incorporate feedback from real-world deployments.

   - **Challenges:** Managing catastrophic forgetting, where the model loses performance on previously learned tasks while adapting to new data. Solutions include multi-task learning or freezing certain layers during updates.

The methodologies for training and fine-tuning LLMs form the backbone of their performance in financial applications. Training focuses on building a strong foundation through diverse datasets and efficient computation, while fine-tuning ensures the model is tailored to specific tasks and domains. By employing structured techniques and iterative refinement, these processes help create models that are not only accurate and efficient but also adaptable to the dynamic nature of the financial sector.

Adaptive Strategies for Domain-Specific Challenges

Fine-tuning large language models (LLMs) for financial applications involves overcoming domain-specific challenges, such as regulatory compliance, sensitive data handling, and the high stakes of financial decision-making. The financial sector presents unique complexities, requiring models to operate with precision, interpretability, and adherence to ethical and legal standards. Tailored strategies are essential to address these challenges and enable the effective deployment of LLMs in critical applications like fraud detection, credit scoring, and portfolio optimization.

Fraud Detection: Identifying Anomalies in Transaction Data

Fraud detection is a high-priority application for LLMs in the financial sector. The goal is to identify fraudulent activities in transactional data, which often exhibit subtle patterns or anomalies. Fine-tuning strategies for this task include

1. **Utilizing Labeled Datasets of Fraudulent Activities:**

    - Collect datasets that include both fraudulent and legitimate transactions, ensuring comprehensive coverage of potential fraud scenarios.

    - Example: Data from credit card networks, e-commerce platforms, and banking institutions labeled with fraud indicators.

2. **Feature Engineering and Embedding Representations:**

    - Integrate domain-specific features, such as transaction amounts, merchant categories, and geolocations, into the model's input.

    - Enhance embeddings by encoding relationships between account holders, transaction timelines, and historical behaviors.

3. **Fine-Tuning with Anomaly Detection Techniques:**

    - Apply methods like supervised learning on labeled fraud datasets or unsupervised learning for anomaly detection when labeled data is sparse.

    - Incorporate autoencoders or Variational Autoencoders (VAEs) to detect deviations from normal patterns.

CHAPTER 3   TRAINING AND FINE-TUNING LLMS

4. **Challenges and Solutions:**

    - **Imbalanced Data:** Fraudulent transactions often form a small fraction of the dataset. Address this with oversampling methods like SMOTE or by applying cost-sensitive learning.

    - **Evolving Fraud Patterns:** Continuously update and fine-tune the model using active learning techniques to adapt to emerging fraud tactics.

Credit Scoring: Analyzing Credit Histories for Risk Assessment

Credit scoring is another critical application, where LLMs assess an individual's creditworthiness based on financial data. Fine-tuning strategies for this task emphasize interpretability and compliance with regulations, such as the Equal Credit Opportunity Act (ECOA).

1. **Adapting Models to Analyze Credit Histories:**

    - Use domain-specific datasets, including credit bureau reports, loan repayment histories, and income data.

    - Incorporate time-series data processing techniques to analyze sequential credit activities, such as missed payments or sudden spikes in credit utilization.

2. **Leveraging Financial Reports and Risk Metrics:**

    - Fine-tune models using structured data extracted from financial statements and risk assessment tools.

    - Map quantitative indicators like debt-to-income ratios, credit utilization rates, and payment delinquencies to model features.

3. **Ensuring Interpretability:**

    - Use explainable AI (XAI) techniques to clarify how predictions are made, ensuring transparency for regulatory compliance and customer trust.

    - Example: LIME (Local Interpretable Model-agnostic Explanations) or SHAP (SHapley Additive exPlanations) can highlight key factors influencing a credit score.

4. **Challenges and Solutions:**

    - **Bias in Data:** Historical biases in credit data can lead to discriminatory scoring. Counteract this by de-biasing datasets and applying fairness constraints during training.

    - **Compliance with Regulations:** Design models to align with regulatory requirements by incorporating auditable processes and maintaining detailed documentation of model decisions.

Portfolio Optimization: Suggesting Investment Strategies

Portfolio optimization involves evaluating and suggesting optimal investment strategies to maximize returns while minimizing risks. Fine-tuning LLMs for this application requires advanced methods to process financial market data and historical trends.

1. **Training Models on Historical Market Trends:**

    - Use datasets comprising stock prices, bond yields, asset allocation strategies, and macroeconomic indicators over time.

    - Incorporate time-series modeling techniques to analyze patterns, seasonality, and correlations between financial instruments.

2. **Evaluating Investment Strategies:**

    - Fine-tune models to generate personalized portfolio suggestions based on an investor's goals, risk tolerance, and financial constraints.

    - Example: Reinforcement learning frameworks can simulate investment scenarios to find optimal strategies under varying market conditions.

3. **Risk Management Integration:**

    - Embed risk assessment metrics such as Value at Risk (VaR), Sharpe ratio, and beta coefficients into the optimization process.

    - Train models to dynamically adjust portfolios in response to market volatility, geopolitical events, or regulatory changes.

CHAPTER 3   TRAINING AND FINE-TUNING LLMS

4. **Challenges and Solutions:**

   - **Data Volatility:** Market data is inherently volatile and may lead to overfitting. Regularize the model using dropout and early stopping during fine-tuning.

   - **Real-Time Adaptation:** Enable real-time updates by integrating streaming financial data using tools like Apache Kafka for continuous model refinement.

Fine-tuning LLMs for domain-specific challenges in finance requires targeted strategies to handle sensitive data, comply with regulations, and address high-stakes tasks. By customizing models for applications like fraud detection, credit scoring, and portfolio optimization, financial institutions can unlock the potential of LLMs to improve decision-making, enhance security, and deliver innovative solutions. These adaptive methodologies ensure that LLMs are not only accurate but also robust and aligned with the complex demands of the financial sector.

Data Preparation for LLMs

Data preparation is the foundation of training and fine-tuning large language models (LLMs). The quality and structure of data directly influence a model's ability to learn meaningful patterns and make accurate predictions. In the context of LLMs, data preparation encompasses collecting, cleaning, augmenting, and formatting datasets to optimize model performance. For financial applications, where precision and domain specificity are paramount, the data preparation process must also address compliance, security, and ethical considerations.

A well-structured data pipeline is fundamental to the success of LLM training and fine-tuning in the financial sector. Each stage of the pipeline contributes to data quality, consistency, and relevance—factors that are especially critical in regulated environments with high sensitivity to data integrity. The following figure outlines a typical end-to-end data pipeline tailored for financial applications, beginning with data collection and culminating in integration with model training workflows.

In Figure 3-3, the pipeline consists of five key stages: data collection, data cleaning, pre-processing, augmentation, and training integration. Each component ensures that raw financial data is transformed into high-quality, structured input suitable for large language model training and domain-specific fine-tuning.

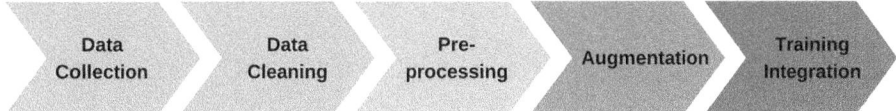

Figure 3-3. *Data Pipeline for Financial Applications*

Data Collection

The first step in preparing data for LLMs is sourcing high-quality and relevant datasets. For financial applications, this requires accessing diverse data types from reliable sources.

- **Sources of Financial Data:**
  - **Structured Data:**
    - Credit reports, transaction records, loan histories, balance sheets, and profit-and-loss statements.
    - Publicly available databases such as World Bank Open Data, SEC filings, and IMF reports.
  - **Unstructured Data:**
    - Financial news articles, market research reports, analyst opinions, and earnings call transcripts.
    - Social media posts and sentiment analysis data for real-time market insights.
  - **External Contextual Data:**
    - Macroeconomic indicators, geopolitical news, and regulatory updates to provide broader context.
- **Diversity of Data Modalities:**
  - Ensure datasets include both text and numerical data to capture the multi-faceted nature of financial analysis.
  - Include multilingual datasets for global financial applications.

- **Ethical and Legal Compliance:**
  - Adhere to data protection regulations like GDPR and CCPA.
  - Anonymize sensitive data to ensure privacy and mitigate risks of misuse.

Data Cleaning

Raw data is often noisy, incomplete, or inconsistent. Cleaning the data ensures it is accurate, relevant, and structured appropriately for LLM training.

- **Handling Missing Values:**
  - For structured data, impute missing values using statistical methods (mean, median) or predictive modeling.
  - For unstructured data, remove incomplete records or flag them for further review.
- **Standardizing Formats:**
  - Normalize numerical data (e.g., currency conversion, interest rates) to a consistent scale.
  - Standardize textual data, ensuring uniformity in formatting and terminology (e.g., "USD" vs. "US Dollars").
- **Tokenization and Text Preprocessing:**
  - Split text into tokens meaningful to the financial domain (e.g., "interest rate" as a single entity).
  - Remove unnecessary elements like stop words, unless they carry domain-specific significance.
- **Outlier Detection and Removal:**
  - Identify outliers in numerical datasets using statistical methods like interquartile range (IQR) or z-scores.
  - For text, remove records with anomalies such as repetitive phrases or unrelated content.

- **Noise Reduction:**

  - Remove irrelevant data, such as advertisements in web-scraped financial news or disclaimers in reports.

  - Use language detection tools to eliminate non-relevant languages from multilingual datasets.

Data Augmentation

Data augmentation expands the dataset to improve model robustness and mitigate class imbalances. This step is particularly useful in financial tasks where labeled data may be limited.

- **Synthetic Data Generation:**

  - Generate artificial data samples using techniques like Variational Autoencoders (VAEs) or Generative Adversarial Networks (GANs).

  - **Example:** Simulate rare events like economic recessions or fraud cases for better model generalization.

- **Paraphrasing and Text Augmentation:**

  - Create variations of textual data while preserving meaning, using models like GPT for paraphrasing.

  - **Example:** Reword financial news headlines to increase diversity in training data.

- **Oversampling and Undersampling:**

  - Balance datasets by oversampling minority classes (e.g., rare defaults) or undersampling majority classes.

  - Use techniques like SMOTE (Synthetic Minority Over-sampling Technique) to generate synthetic examples for underrepresented classes.

- **Feature Engineering:**
  - Derive new features from existing data, such as calculating credit utilization rates from loan and income data.
  - Create composite features like risk scores or sentiment indices from combined datasets.

Dataset Splitting

Properly dividing the dataset into training, validation, and testing subsets ensures that the model is both accurate and generalizable.

- **Standard Splitting Ratios:**
  - Use a typical split of 70% for training, 20% for validation, and 10% for testing.
  - For imbalanced datasets, stratify the splits to maintain proportional representation of classes.
- **Cross-Validation:**
  - Employ k-fold cross-validation for tasks with limited data to evaluate model performance across multiple subsets.
  - Example: In a fraud detection task, validate the model on different folds to ensure consistent performance.
- **Holdout Data for Final Testing:**
  - Reserve a completely unseen dataset for final evaluation to prevent overfitting and assess real-world performance.

Data Governance and Documentation

Financial data often involves compliance and audit requirements. Proper governance and documentation are essential for transparency and accountability.

- **Metadata Management:**
  - Maintain detailed metadata for each dataset, including its source, date of collection, and preprocessing steps applied.

- **Version Control:**
  - Use tools like DVC (Data Version Control) to track changes in datasets and ensure reproducibility of experiments.
- **Data Lineage:**
  - Document the flow of data from its source to its final processed state to facilitate auditing and compliance checks.

Tools and Frameworks for Data Preparation

Modern tools streamline the data preparation process, improving efficiency and scalability.

- **Data Cleaning and Transformation:**
  - **Tools:** Pandas, PySpark, OpenRefine
  - **Example:** Use Pandas for quick exploratory cleaning and PySpark for large-scale transformations.
- **Data Augmentation:**
  - **Tools:** NLTK, spaCy, GAN-based frameworks
  - **Example:** Use NLTK for basic text augmentation and GANs for synthetic data generation.
- **Data Validation:**
  - **Tools:** Great Expectations, TensorFlow Data Validation
  - **Example:** Validate schema consistency and check for missing values automatically.

Data preparation for LLMs is a meticulous process that sets the stage for successful training and fine-tuning. In the financial sector, where data is diverse, sensitive, and highly regulated, a robust data preparation pipeline ensures that models are accurate, interpretable, and aligned with compliance requirements. By investing in well-curated and properly processed datasets, organizations can unlock the full potential of LLMs, driving innovation and precision in financial applications.

Training Methodologies

Training large language models (LLMs) is a complex and resource-intensive process that requires careful orchestration of data, algorithms, and computational resources. For financial applications, training methodologies must also consider domain-specific challenges such as precision, interpretability, and compliance. This section delves into the critical elements of training LLMs, highlighting strategies and techniques that optimize performance, scalability, and adaptability.

Pretraining: Building a Foundational Model

Pretraining is the first phase of training LLMs, where the model learns general language patterns from large, diverse datasets.

- **Purpose of Pretraining:**
  - To equip the model with a foundational understanding of language syntax, semantics, and relationships.
  - Pretraining creates a versatile base that can be fine-tuned for specific financial tasks, such as fraud detection or risk assessment.

- **Dataset Requirements:**
  - Use large-scale datasets like Common Crawl, Wikipedia, and financial news archives.
  - Incorporate domain-specific data, such as regulatory filings and market analysis reports, for better alignment with financial terminology.

- **Techniques for Efficient Pretraining:**
  - **Masked Language Modeling (MLM):** Used in models like BERT, where the model predicts masked words in a sentence to learn contextual representations.
  - **Causal Language Modeling (CLM):** Used in autoregressive models like GPT, where the model predicts the next word based on preceding words.

- **Challenges and Solutions:**
  - **High Computational Costs:** Mitigate by using distributed training techniques and efficient hardware like TPUs or GPUs.
  - **Data Quality Variability:** Address by preprocessing and curating datasets to remove noise and irrelevant content.

Distributed Training: Scaling for Large Models

Modern LLMs often contain billions of parameters, necessitating distributed training techniques to handle computational and memory demands.

- **Types of Distributed Training:**
  - **Data Parallelism:** Splits data across multiple devices, with each device processing a unique data shard while maintaining identical model copies.
  - **Model Parallelism:** Splits the model across multiple devices, enabling the training of large models that exceed the memory capacity of a single device.
  - **Pipeline Parallelism:** Breaks the training process into stages, each handled by a different device, optimizing resource utilization.
- **Frameworks for Distributed Training:**
  - TensorFlow Distributed, PyTorch Distributed, DeepSpeed, and Horovod.
  - Example: DeepSpeed enables memory-efficient training of models with tens of billions of parameters.

Best Practices:
- Use gradient accumulation to simulate large batch sizes on memory-constrained hardware.
- Implement mixed-precision training to reduce memory usage and improve computation speed without sacrificing accuracy.

Optimization Algorithms

Optimization algorithms play a central role in adjusting the model's weights to minimize the loss function during training.

- **Gradient-Based Optimization:**
  - Commonly used algorithms include Stochastic Gradient Descent (SGD), Adam, and RMSProp.
  - AdamW, a variant of Adam, incorporates weight decay for better regularization.
- **Learning Rate Schedules:**
  - Gradual warm-up at the beginning of training followed by decay schedules (e.g., cosine annealing or exponential decay) ensures stable convergence.
  - Cyclical learning rates can help escape local minima and improve generalization.
- **Loss Functions:**
  - Use cross-entropy loss for classification tasks and mean squared error (MSE) for regression tasks.
  - For imbalanced datasets, apply weighted loss functions to penalize errors on underrepresented classes more heavily.

Regularization Techniques

Regularization prevents overfitting, ensuring that the model generalizes well to unseen data.

- **Dropout:**
  - Randomly deactivates a fraction of neurons during training to prevent co-adaptation.
  - Typical dropout rates range from 0.1 to 0.5, depending on the model architecture.

- **Weight Decay:**
  - Adds a penalty to large weight values, encouraging simpler models that generalize better.
- **Data Augmentation:**
  - Introduce variability in training data through techniques like paraphrasing and synthetic data generation.
  - Example: Generate synthetic financial statements to enrich datasets for rare scenarios.

Checkpointing and Logging

Checkpointing and logging are essential for monitoring training progress and ensuring recoverability.

- **Checkpointing:**
  - Save model weights at regular intervals to prevent loss of progress due to interruptions.
  - Maintain multiple checkpoints to revert to earlier stages if training diverges.
- **Logging Metrics:**
  - Track key performance indicators like training loss, validation accuracy, and resource utilization.
  - Use tools like TensorBoard or MLflow for real-time visualization of training metrics.

Handling Data and Model Bias

Bias in data or model predictions can lead to unfair or inaccurate outcomes, particularly in high-stakes applications like credit scoring.

- **Bias Mitigation Techniques:**
  - Use de-biasing algorithms during data preprocessing to remove systemic patterns that disadvantage certain groups.
  - Apply sample reweighting, where higher weights are assigned to underrepresented or disadvantaged groups to balance the training distribution.
  - Leverage adversarial de-biasing, where a secondary model is trained to detect sensitive attributes from the model's internal representations—and the primary model is penalized if such attributes are predictable, promoting fairness in learned representations.
  - Regularly audit model predictions using fairness metrics (e.g., equalized odds, demographic parity) to identify and address emerging biases.
- **Fairness Constraints:**
  - Incorporate fairness-aware loss functions into model training to ensure equitable treatment across protected attributes (e.g., gender, age, ethnicity).
  - These constraints help optimize for both performance and fairness, ensuring that models do not unintentionally prioritize accuracy at the cost of discriminatory outcomes.

Challenges in Training LLMs

Training LLMs is fraught with challenges that require targeted solutions:

- **High Computational Costs:**
  - Use cloud-based services or distributed systems to reduce infrastructure constraints.
  - Optimize memory usage with techniques like gradient checkpointing.

- **Model Convergence Issues:**
  - Employ advanced initialization techniques to stabilize training, such as LayerNorm or Xavier initialization.
- **Data Diversity and Representation:**
  - Ensure datasets are representative of the financial domain to reduce domain gaps.

Training large language models is a critical process that requires a balanced approach to data, computation, and algorithmic design. By leveraging advanced training methodologies like distributed training, optimization algorithms, and regularization techniques, practitioners can build robust models that meet the stringent demands of financial applications. Thoughtful implementation of these methodologies ensures that models are not only accurate but also efficient and adaptable to the dynamic nature of the financial sector.

Fine-Tuning Techniques

Fine-tuning is the process of adapting a pretrained large language model (LLM) to a specific domain or task by updating its parameters with task-specific data. In the financial sector, fine-tuning enables LLMs to handle nuanced requirements such as fraud detection, credit scoring, and portfolio optimization while maintaining high accuracy, compliance, and interpretability. This section explores advanced fine-tuning techniques that maximize model performance and ensure alignment with the specialized needs of financial applications.

Task-Specific Fine-Tuning

Task-specific fine-tuning involves customizing the LLM to perform a clearly defined task, such as classification, regression, or sequence generation.

- **Key Components of Task-Specific Fine-Tuning:**
  - **Dataset Preparation:** Use labeled datasets relevant to the task, such as transaction records labeled for fraud or customer credit histories annotated with risk scores.

- **Training Objective:** Choose the appropriate loss function, such as cross-entropy for classification tasks or mean squared error for regression.

- **Architecture Modification:** Add task-specific layers (e.g., classification heads) on top of the pretrained model.

- **Example Applications:**

  - **Fraud Detection:** Fine-tune the model on transactional data to identify anomalous patterns indicative of fraud.

  - **Sentiment Analysis:** Adapt the model to analyze market sentiment from financial news or social media.

Domain-Adaptive Pretraining (DAPT)

DAPT is an intermediate step between general pretraining and task-specific fine-tuning. It involves pretraining the model further on domain-specific corpora to better align it with the language and context of a particular field.

- **Steps in DAPT:**

  - **Domain-Specific Data Collection:** Gather large volumes of unannotated text from financial sources such as regulatory filings, analyst reports, and earnings call transcripts.

  - **Training Objective:** Use masked language modeling (MLM) or causal language modeling (CLM) to continue pretraining the model on domain-specific data.

- **Benefits:**

  - Improves the model's understanding of domain-specific terminology, such as "collateralized debt obligations" or "credit default swaps."

  - Reduces the data requirement for subsequent task-specific fine-tuning by aligning the model with the domain context.

Transfer Learning

Transfer learning leverages knowledge from a pretrained model to improve performance on a related but distinct task. In fine-tuning, transfer learning involves reusing the general linguistic capabilities of an LLM and adapting it to specialized tasks.

- **Approaches to Transfer Learning:**
  - **Feature-Based Transfer Learning:** Extract embeddings from the pretrained model and use them as input features for a downstream task.
  - **Fine-Tuning Transfer Learning:** Update all or selected layers of the pretrained model using task-specific data.
- **Advantages:**
  - Saves computational resources compared to training a model from scratch.
  - Enables rapid adaptation to tasks with limited labeled data.

Few-Shot and Zero-Shot Learning

Few-shot and zero-shot learning are advanced fine-tuning techniques that reduce the reliance on large labeled datasets.

- **Few-Shot Learning:**
  - Fine-tune the model using a small number of labeled examples.
  - **Example:** Train a model to classify fraudulent transactions using only a few hundred labeled examples.
- **Zero-Shot Learning:**
  - Leverage the model's inherent knowledge to perform tasks without task-specific fine-tuning.
  - **Example:** Use a pretrained LLM to summarize financial news without additional training.

- **Benefits:**
  - Reduces the cost and effort associated with data labeling.
  - Allows quick adaptation to new tasks.

Hyperparameter Tuning During Fine-Tuning

Hyperparameter tuning involves optimizing parameters that control the training process to improve model performance.

- **Key Hyperparameters to Optimize:**
  - **Learning Rate:** A low learning rate prevents catastrophic forgetting of pretrained knowledge. Use warm-up schedules followed by decay.
  - **Batch Size:** Larger batch sizes improve gradient stability but require more memory. Use dynamic batching when resource-constrained.
  - **Dropout Rate:** Regularize the model to prevent overfitting, particularly when fine-tuning on small datasets.
- **Optimization Techniques:**
  - Use grid search, random search, or advanced methods like Bayesian optimization to identify the best hyperparameter combinations.

Regularization Techniques

Regularization ensures that the fine-tuned model generalizes well to unseen data, reducing the risk of overfitting.

- **Layer Freezing:**
  - Freeze the lower layers of the model to retain general linguistic knowledge while fine-tuning only the upper layers for task-specific adaptations.

- Example: Freeze the transformer layers and fine-tune the classification head for a credit scoring task.
- **Dropout and Weight Regularization:**
  - Use dropout to deactivate a fraction of neurons during training, promoting robust learning.
  - Apply weight decay to discourage large parameter values that could lead to overfitting.

Multi-Task Fine-Tuning

Multi-task fine-tuning involves training the model simultaneously on multiple related tasks, improving its ability to generalize and perform across domains.

- **Implementation:**
  - Use a shared encoder-decoder architecture where the encoder processes input for all tasks, and task-specific decoders handle task outputs.
  - **Example:** Train the model on fraud detection and credit scoring tasks concurrently.
- **Advantages:**
  - Enables the model to learn shared representations that benefit all tasks.
  - Reduces the cost and time associated with fine-tuning separate models for each task.

Active Fine-Tuning

Active fine-tuning integrates human feedback into the fine-tuning process, ensuring that the model aligns closely with business objectives and regulatory requirements.

- **Human-in-the-Loop Systems:**
  - Use expert feedback to validate and refine model predictions during training.

- **Example:** Financial analysts review credit risk scores generated by the model and provide corrective input.

- **Reinforcement Learning from Human Feedback (RLHF):**
  - Train the model using a reward system based on human feedback to align its outputs with desired outcomes.
  - **Example:** Reward the model for generating risk reports that comply with regulatory guidelines.

Handling Domain Shifts

Domain shifts occur when the model is exposed to new data distributions that differ from the fine-tuning dataset. Strategies include

- **Incremental Fine-Tuning:**
  - Periodically fine-tune the model on updated datasets to align with changing market conditions or regulatory frameworks.

- **Continuous Learning Pipelines:**
  - Implement pipelines that automate data collection, preprocessing, and fine-tuning to keep the model up-to-date.

Evaluation During Fine-Tuning

Rigorous evaluation ensures that the fine-tuned model meets the desired performance and compliance standards.

- **Key Metrics:**
  - Use task-specific metrics such as F1-score, ROC-AUC, or RMSE to measure performance.
  - Employ fairness and bias metrics to ensure equitable predictions.

- **Validation and Testing:**
  - Use a validation dataset during fine-tuning to monitor performance.
  - Test on out-of-sample data to assess generalization capabilities.

Fine-tuning is a critical step in adapting LLMs to the specialized needs of financial applications. By employing advanced techniques such as domain-adaptive pretraining, multi-task fine-tuning, and active learning, practitioners can create models that are accurate, robust, and aligned with business objectives. These fine-tuning strategies not only optimize performance but also address the unique challenges of the financial domain, enabling LLMs to deliver high-impact solutions with precision and reliability.

Challenges in Training and Fine-Tuning

Training and fine-tuning large language models (LLMs) for financial applications come with significant challenges due to the complexity of the domain, the need for high accuracy, and strict compliance requirements. These challenges can impact both the model's performance and its adoption in real-world applications. Key challenges include

Data-Related Challenges

- **Data Scarcity in Specific Use Cases:**

    Financial datasets for specialized tasks, such as fraud detection or rare credit risk events, are often limited in size.

    - **Solution:** Employ data augmentation techniques and leverage synthetic data generation using Variational Autoencoders (VAEs) or Generative Adversarial Networks (GANs).

- **Data Sensitivity and Privacy:**

    Financial data is sensitive and often subject to strict privacy regulations like GDPR and CCPA.

    - **Solution:** Use anonymization techniques, differential privacy, and secure multiparty computation to protect sensitive information.

- **Data Bias and Imbalance:**

  Historical biases in financial datasets can lead to discriminatory outcomes, while imbalanced datasets can cause models to underperform on minority classes.

  - **Solution:** Apply fairness-aware algorithms, stratified sampling, and reweighting techniques to correct for biases.

Computational Challenges

- **High Resource Requirements:**

  Training large-scale LLMs demands substantial computational resources, often requiring access to distributed systems with GPUs or TPUs.

  - **Solution:** Optimize training through distributed strategies like data and model parallelism, and utilize frameworks like DeepSpeed or Horovod for scalability.

- **Model Convergence Issues:**

  Training can suffer from vanishing gradients, overfitting, or underfitting, leading to suboptimal model performance.

  - **Solution:** Regularize training with dropout, weight decay, and batch normalization. Employ techniques like early stopping and learning rate schedules to ensure convergence.

Domain-Specific Challenges

- **Regulatory Compliance:**

  Financial models must align with strict regulatory standards to avoid legal repercussions.

  - **Solution:** Integrate compliance checks during data preparation and model evaluation stages, and document all training and fine-tuning processes for auditability.

- **Evolving Financial Landscapes:**

  Models need to adapt to dynamic market conditions and new regulations.

  - **Solution:** Implement continuous learning pipelines that periodically update the model with new data to address evolving requirements.

Best Practices for Training and Fine-Tuning

Training and fine-tuning large language models (LLMs) are resource-intensive and complex processes that require careful planning and execution to maximize efficiency and effectiveness. The financial domain adds further challenges, such as regulatory compliance, sensitivity to errors, and the dynamic nature of the industry. Adhering to best practices ensures that models are not only accurate but also robust, interpretable, and aligned with industry requirements. Below is a detailed exploration of best practices for each phase of training and fine-tuning.

Data Preparation

High-quality data is foundational to effective training and fine-tuning. Models trained on well-curated datasets exhibit better generalization and performance.

- **Curate Domain-Specific Datasets:**
  - Use data sources relevant to the financial domain, such as credit histories, transaction records, market reports, and regulatory filings.
  - Ensure data is representative of the tasks the model will perform, such as fraud detection, credit scoring, or market prediction.
- **Handle Data Imbalances:**
  - Financial datasets often exhibit class imbalance (e.g., rare fraudulent transactions). Address this by employing techniques such as SMOTE (Synthetic Minority Over-sampling Technique), undersampling, or cost-sensitive learning.

- **Implement Data Preprocessing Pipelines:**
  - Clean the data to remove noise, duplicates, and irrelevant information.
  - Tokenize and normalize textual data to standardize input formats for the model.
  - For numerical data, scale features using methods like min-max scaling or z-score normalization.
- **Augment Data for Robustness:**
  - Use synthetic data generation to enrich datasets, especially for rare events like defaults or fraud.
  - Apply paraphrasing techniques to expand textual data diversity without altering meaning.

Training Best Practices

Training LLMs involves optimizing the model on large-scale datasets to develop generalized language understanding.

- **Optimize Hyperparameters:**
  - Tune critical parameters such as learning rate, batch size, weight decay, and dropout rate to achieve optimal performance.
  - Use adaptive learning rate schedules like cosine annealing or warm-up followed by exponential decay.
- **Leverage Distributed Training:**
  - Use hybrid parallelism (data, model, and pipeline parallelism) to distribute computational load across multiple GPUs or TPUs.
  - Tools like DeepSpeed, Horovod, or PyTorch Distributed reduce training time and improve scalability.
- **Use Pretraining with Diverse Datasets:**
  - Pretrain models on large, heterogeneous datasets to develop a robust understanding of language patterns.

- For finance-specific applications, consider pretraining on datasets with domain-specific terminologies, such as regulatory filings or market analysis reports.

- **Regularize Training:**
  - Prevent overfitting by using regularization techniques such as dropout, early stopping, and weight decay.
  - Augment the training process with noise injection or adversarial training to improve model resilience.

- **Implement Checkpointing and Logging:**
  - Save model states at regular intervals to ensure recoverability in case of interruptions.
  - Use logging frameworks like MLflow to monitor metrics such as loss, accuracy, and validation performance.

Fine-Tuning Best Practices

Fine-tuning adapts a pretrained model to specific tasks or domains, such as credit scoring, portfolio optimization, or risk assessment.

- **Domain-Adaptive Pretraining (DAPT):**
  - Before fine-tuning, continue pretraining the model on domain-specific datasets to align its language understanding with financial terminology and concepts.
  - **Example:** Pretrain on datasets like SEC filings or financial reports before fine-tuning for credit risk assessment.

- **Task-Specific Fine-Tuning:**
  - Use task-specific datasets to fine-tune the model for applications like fraud detection or market prediction.
  - Apply transfer learning techniques to leverage the knowledge acquired during pretraining.

- **Iterative Fine-Tuning:**
  - Continuously fine-tune the model as new data becomes available to keep it updated with evolving market conditions or regulatory requirements.
  - Implement active learning pipelines to integrate user feedback and improve the model iteratively.
- **Regularization During Fine-Tuning:**
  - Prevent overfitting by freezing certain layers of the model during fine-tuning, especially those learned during pretraining.
  - Use techniques like weight regularization or multi-task learning to maintain generalization capabilities.

Evaluation and Monitoring

Rigorous evaluation and monitoring ensure that the model performs reliably and meets predefined success criteria.

- **Establish Performance Metrics:**
  - Use domain-specific metrics like F1-score, ROC-AUC, and precision-recall for fraud detection.
  - For regression tasks, evaluate performance using RMSE (Root Mean Square Error) or MAPE (Mean Absolute Percentage Error).
- **Conduct Robust Testing:**
  - Test models on out-of-sample datasets to ensure they generalize well to unseen data.
  - Use adversarial examples to test the robustness of predictions in edge cases.
- **Integrate Explainability:**
  - Employ tools like SHAP (SHapley Additive exPlanations) or LIME (Local Interpretable Model-agnostic Explanations) to provide interpretability for model predictions.

- Ensure explainability aligns with regulatory requirements, particularly in high-stakes applications like credit scoring.

- **Monitor for Model Drift:**
  - Use statistical methods to detect shifts in data distribution that could impact model performance.
  - Schedule periodic retraining or fine-tuning to address drift.

- **Enhance Efficiency with Quantized Fine-Tuning:**
  - Use Q-LoRA (Quantized Low-Rank Adaptation) to fine-tune large models with lower memory and compute overhead, making them deployable in resource-constrained environments.
  - Apply GPTQ (Post-Training Quantization) to compress models down to 8-bit or lower precision, significantly reducing latency and improving inference speed—without major performance trade-offs.

Compliance and Ethics

Financial models must adhere to ethical guidelines and comply with regulatory standards.

- **Ensure Data Privacy and Security:**
  - Use anonymization techniques, secure data storage, and encryption to protect sensitive financial data.
  - Implement differential privacy mechanisms to add noise to datasets without compromising accuracy.

- **Maintain Audit Trails:**
  - Document the training and fine-tuning process, including dataset sources, preprocessing steps, and hyperparameter configurations.
  - Create reproducible workflows for model updates and evaluations.

- **Embed Fairness and Bias Mitigation:**
  - Regularly audit models for biases in predictions or decision-making.
  - Use fairness constraints and post-processing techniques to correct for any detected bias.

Adhering to these best practices ensures that LLMs are trained and fine-tuned efficiently and effectively for financial applications. By focusing on data quality, optimization techniques, and domain-specific adaptation, practitioners can build robust, scalable, and compliant models. These practices are not only critical for achieving technical excellence but also for fostering trust and transparency in financial decision-making systems.

Case Study: Fine-Tuning for Risk Management
Context and Problem Statement

Effective risk management is a cornerstone of the financial sector, enabling institutions to predict and mitigate risks such as loan defaults, credit downgrades, and market instability. Traditional risk models rely heavily on structured data and statistical methods, which may fail to capture complex patterns or adapt to evolving scenarios. A multinational financial institution sought to enhance its risk management capabilities by fine-tuning a large language model (LLM) for predictive analysis and decision-making.

The primary goal was to develop an LLM that could analyze diverse data sources, including credit reports, financial news, and macroeconomic indicators, to assess and predict risk profiles with high accuracy and interpretability.

Objective

The institution aimed to achieve the following:

1. Enhance the accuracy of risk predictions.
2. Ensure the model can handle diverse data modalities (structured and unstructured).

3. Maintain regulatory compliance and explainability in predictions.

4. Adapt the model to evolving financial conditions and regulatory requirements.

Methodology

The process involved four key phases: data preparation, fine-tuning, evaluation, and deployment.

1. **Data Preparation**

   The success of fine-tuning depends heavily on the quality and relevance of the data. For this project:

   - **Data Collection:**
     - **Structured Data:** Historical credit scores, repayment histories, and default rates.
     - **Unstructured Data:** Financial news articles, analyst reports, and regulatory filings.
     - **External Indicators:** Macroeconomic metrics like GDP growth rates, interest rates, and inflation levels.

   - **Data Cleaning and Preprocessing:**
     - Removed noise, duplicates, and irrelevant information.
     - Tokenized unstructured text data while preserving domain-specific terms such as "collateralized debt" or "credit default swap."
     - Normalized numerical data (e.g., debt-to-income ratios) for consistency.

   - **Data Augmentation:**
     - Synthetic data generation was used to simulate rare events, such as economic recessions or large-scale defaults.
     - Oversampled datasets for minority classes, like high-risk customers, to balance the training data.

2. **Fine-Tuning Process**

    - **Domain-Adaptive Pretraining (DAPT):**
        - The base LLM was pretrained further on domain-specific datasets, such as regulatory filings, to familiarize it with financial terminology and context.
        - This stage used large-scale corpora of financial documents to enhance the model's understanding of risk-related language.

    - **Task-Specific Fine-Tuning:**
        - The model was fine-tuned on a labeled dataset of customer profiles, transaction histories, and credit events (e.g., defaults and delinquencies).
        - **Objective Function:** Minimized a weighted loss function to address class imbalances, ensuring accurate predictions for both high-risk and low-risk categories.

    - **Hyperparameter Optimization:**
        - Grid search was used to identify optimal values for learning rate, batch size, and dropout rate.
        - Early stopping was employed to prevent overfitting during fine-tuning.

3. **Evaluation**

    Rigorous evaluation ensured the fine-tuned model met accuracy, interpretability, and compliance requirements.

    - **Performance Metrics:**
        - Accuracy, F1-score, and ROC-AUC were used to measure prediction quality.
        - Precision-recall analysis helped assess the model's ability to identify high-risk cases without excessive false positives.

- **Interpretability:**
  - Explainability tools like SHAP (SHapley Additive exPlanations) were integrated to highlight the factors influencing risk scores.
  - For example, a SHAP analysis revealed that a sharp increase in credit utilization was a key driver for high-risk predictions.
- **Compliance Checks:**
  - The model was audited to ensure it aligned with regulatory requirements like the Equal Credit Opportunity Act (ECOA).
  - Bias testing confirmed that predictions were not influenced by protected attributes such as race, gender, or age.

4. **Deployment and Monitoring**
   - **Integration with Existing Systems:**
     - The model was deployed into the institution's risk assessment pipeline, interfacing with loan approval systems and credit monitoring tools.
     - APIs were developed to allow real-time risk scoring based on incoming data.
   - **Continuous Monitoring:**
     - Model drift was monitored by comparing predictions with actual outcomes.
     - A retraining pipeline was established to update the model periodically with new data, ensuring it adapted to changing market conditions.

Results

Improved Accuracy:
- The fine-tuned model achieved a 20% improvement in predictive accuracy compared to the institution's traditional risk models.
- High-risk cases were identified with greater precision, reducing false positives by 15%.

Enhanced Interpretability:

- Stakeholders gained confidence in the model's predictions through detailed explanations of key risk factors.
- Compliance officers were able to audit and validate the model's decision-making process.

Scalability and Adaptability:

- The model seamlessly scaled to handle large datasets from multiple sources.
- Continuous updates ensured it remained effective despite evolving financial trends and regulatory changes.

Lessons Learned

Importance of Data Diversity:

- Combining structured and unstructured data provided a holistic view of risk factors, enhancing model performance.

Explainability as a Priority:

- Transparent predictions were crucial for stakeholder trust and regulatory approval.

Ongoing Adaptation:

- Establishing a retraining pipeline ensured the model could respond dynamically to new risks and market conditions.

This case study illustrates the transformative potential of fine-tuned LLMs in financial risk management. By addressing domain-specific challenges through careful data preparation, rigorous fine-tuning, and robust evaluation, the institution was able to significantly improve its risk prediction capabilities. This success highlights the strategic value of LLMs in advancing decision-making and operational efficiency in the financial sector.

Training and fine-tuning large language models (LLMs) are pivotal steps in transforming these advanced models into practical tools tailored to specific applications. This chapter has detailed the structured methodologies and adaptive techniques required to prepare, train, and fine-tune LLMs effectively, with a special focus on the financial sector. From curating high-quality data to employing advanced optimization and regularization methods, each step plays a critical role in ensuring model accuracy, scalability, and compliance.

In the financial domain, where precision, interpretability, and reliability are paramount, leveraging best practices in training and fine-tuning is essential to unlock the full potential of LLMs. Through fine-tuning techniques such as domain-adaptive pretraining, task-specific adaptation, and multi-task learning, LLMs can address complex challenges like fraud detection, credit risk assessment, and portfolio optimization. These processes enable financial institutions to enhance decision-making, mitigate risks, and maintain regulatory alignment.

As we transition to the next chapter, we move beyond the training phase and explore **deployment strategies for LLMs**. This involves addressing the practical challenges of integrating LLMs into real-world financial systems, optimizing their performance in production environments, and establishing robust monitoring and feedback mechanisms. Chapter 4 will guide you through the critical steps required to deploy LLMs efficiently, ensuring they deliver actionable insights and value in live operational contexts.

Conclusion

Training and fine-tuning large language models (LLMs) are crucial processes that enable these models to handle domain-specific challenges and deliver actionable insights, especially in the highly regulated and dynamic financial sector. This chapter has emphasized the importance of high-quality data as the foundation for success, exploring strategies for data collection, cleaning, and augmentation tailored to financial applications. By adopting structured training methodologies like distributed training and domain-adaptive pretraining, practitioners can equip LLMs with the capabilities to address diverse tasks, from fraud detection to credit risk assessment.

Fine-tuning further customizes these models, ensuring they align with specific use cases while maintaining compliance, interpretability, and scalability. Techniques such as task-specific tuning, regularization, and hyperparameter optimization play pivotal roles in enhancing model performance and robustness. Moreover, continuous learning pipelines and active feedback loops allow LLMs to adapt to evolving market trends and regulatory requirements, ensuring their relevance and reliability in real-world financial applications.

CHAPTER 3 TRAINING AND FINE-TUNING LLMS

With trained and fine-tuned LLMs ready to address specific financial challenges, the next step is deploying these models in live operational environments. Deployment is a critical phase that ensures LLMs can deliver real-time insights, make accurate predictions, and integrate seamlessly into existing financial systems. Chapter 4 focuses on the strategies, tools, and considerations necessary for transitioning LLMs from development to production, with an emphasis on scalability, security, and performance optimization.

CHAPTER 4

Deployment Strategies for LLMs

This chapter explores key strategies for deploying large language models (LLMs) in production environments, particularly within the finance industry. It focuses on the essential components for building efficient, scalable, and reliable deployment systems for LLMs, ensuring that models can handle high-volume, real-time workloads while meeting strict regulatory and performance standards. The chapter also provides best practices for optimizing performance, monitoring system health, and managing resource usage to ensure smooth and cost-effective operations. By understanding how to efficiently manage LLM deployment, organizations can ensure their models deliver accurate and timely results without interruptions.

Structure

This chapter covers the following topics:

- **Deployment Pipelines**: Automating and streamlining model deployment through CI/CD, model versioning, and automated testing.
- **Monitoring and Logging:** Ensuring real-time performance tracking, error detection, and resource optimization.
- **Performance Optimization**: Techniques to scale LLMs, reduce latency, and improve efficiency through caching and load balancing.
- **LLM Ops Lifecycle:** A framework for understanding the development, deployment, and continuous monitoring of LLMs, focusing on regulatory compliance, data security, and system reliability in the finance industry.

CHAPTER 4 DEPLOYMENT STRATEGIES FOR LLMS

Objectives

At the end of this chapter, you will understand how to build and automate deployment pipelines for LLMs in financial environments. You will learn how to set up monitoring and logging systems to ensure ongoing model performance and reliability. Applying performance optimization techniques to maximize the efficiency, scalability, and responsiveness of LLMs in production environments will also be discussed. In this chapter, you will also comprehend the critical stages of the LLM Ops Lifecycle and how they align with the unique challenges of financial systems, ensuring continuous model improvement, regulatory compliance, and risk management.

Introduction

The deployment of large language models (LLMs) in finance has transformed key areas such as fraud detection, customer service, credit scoring, and regulatory compliance. However, deployment is just the beginning—ensuring long-term performance requires continuous monitoring and maintenance. Without a structured approach, models may degrade over time, fail to adapt to changing data patterns, or violate regulatory standards, leading to financial and reputational risks.

This section explores why monitoring and maintaining LLMs is essential for their continued success in financial applications. It highlights common challenges in dynamic environments and provides actionable strategies to address them effectively.

Deployment Pipelines

Efficiently deploying large language models (LLMs) in production environments, especially in high-stakes industries like finance, requires a well-structured and reliable pipeline. A deployment pipeline ensures the seamless transition of models from development to production, facilitating continuous integration, automated testing, version control, and gradual rollouts. By streamlining these processes, organizations can ensure that their LLMs are deployed reliably, with minimal downtime, and are capable of scaling efficiently to handle real-time demands.

In this section, you will explore key components of an effective deployment pipeline: Continuous Integration/Continuous Deployment (CI/CD), model versioning, automated testing, and canary deployments.

Continuous Integration/Continuous Deployment (CI/CD)

The goal of a CI/CD pipeline is to automate the process of deploying new or updated LLMs to production environments with minimal downtime and human intervention. CI/CD practices are foundational to modern deployment pipelines. In the context of LLMs, CI/CD ensures that new models or updates are continuously tested, integrated, and deployed into production without disrupting operations. This automation is crucial in financial systems where real-time operations, such as fraud detection or risk assessment, require high availability and reliability.

Note CI/CD in Finance Requires Extra Security Measures—Since financial applications involve sensitive data, ensure that deployment pipelines are secured with proper authentication, encryption, and access controls.

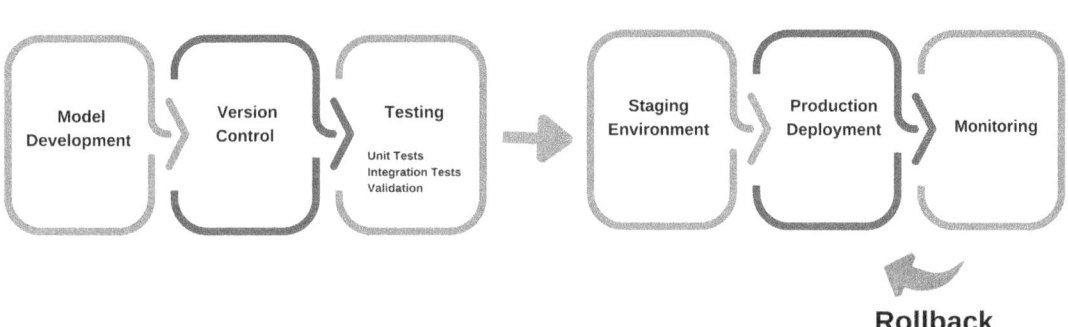

Figure 4-1. *LLM Deployment Pipeline*

As shown in Figure 4-1, the LLM deployment process involves several critical stages, beginning with model development, where the large language model is designed, trained, and refined to meet specific requirements. Version control, utilizing tools like Git, plays a crucial role in tracking changes to the model's code, configurations, and training data, ensuring collaboration and reproducibility throughout the development lifecycle. Rigorous testing is then conducted, encompassing unit tests to verify the functionality of individual components, integration tests to validate the interaction between different parts of the system, and automated validation checks to assess the model's performance, accuracy, and potential biases. Before full-scale deployment, a staging environment, which mirrors the production setup, is used for final testing and quality assurance. Deployment itself is the process of making the LLM operational and accessible for use in a production environment. Post-deployment, monitoring is essential to track the LLM's performance and usage patterns and to identify any potential issues, including logging errors and monitoring key performance metrics. Finally, a rollback mechanism provides a safety net, enabling a quick reversion to a previous stable version of the LLM in the event of problems or failures in the production environment.

CI/CD Process:

In CI, model updates (such as retraining on new data or fine-tuning for specific tasks) are continuously integrated into a shared repository. Each update is automatically tested to verify its functionality and accuracy. This process ensures that models are always up-to-date and compatible with the production environment.

Once a model passes the integration and testing phases, it is automatically deployed to production. By automating this process, financial institutions can avoid lengthy manual processes and reduce the likelihood of errors. With CD, updates can be deployed rapidly, ensuring that the LLMs remain responsive to new data and evolving business needs. Some of the benefits of the CI/CD pipelines are as follows:

1. **Minimal Downtime:** Automating the deployment of LLMs means that updates can occur without interrupting live operations, which is critical in sectors where continuous availability is essential.

2. **Reduced Manual Effort:** Automating repetitive tasks in deployment reduces the need for manual intervention, freeing up resources and reducing the potential for human error.

3. **Rapid Response to Data Changes:** In fast-moving industries like finance, the ability to quickly deploy updated models in response to changing data (e.g., market shifts or new regulatory requirements) is a significant advantage.

Caution Avoid Overfitting Deployment Triggers—Frequent retraining and deployment of models can lead to overfitting if not monitored carefully. Ensure that retrained models genuinely improve performance before pushing them to production.

Model Versioning

The goal is to manage and track multiple versions of models to ensure stability, transparency, and the ability to revert to earlier versions if needed. Model versioning is a critical component of LLM deployment pipelines. As models evolve through retraining and fine-tuning, it is important to maintain a clear record of each version, ensuring that previous models can be recalled or reverted to in the event of an issue with a new deployment. Versioning also facilitates collaboration between teams, allowing data scientists, engineers, and business stakeholders to track model updates and their performance over time.

In practice, as shown in Table 4-1, popular tools used for model versioning include the following:

- **MLflow** (open-source) for model registry and tracking.
- **DVC** (data version control) for tracking code, data, and model files together.
- **Weights and Biases** for capturing and comparing experimentation history.

Table 4-1. Common Tools Used for Model Versioning

Tool	Purpose	Example Use Case
MLflow	Model registry, version tracking	Track and register each LLM fine-tune step with metadata like accuracy, dataset
DVC	Code/data/model version control	Reproduce an older experiment by syncing the exact model + dataset versions
Weights and Biases	Experiment tracking and comparison	Compare 20 different prompt tuning experiments and identify the best-performing one

The key aspects of model versioning are as follows:

1. **Version Control for Models:** Just like software code, LLMs need version control to track changes, manage different iterations, and maintain a history of updates. This ensures that every change—whether an algorithmic adjustment, retraining with new data, or hyperparameter tuning—is logged and can be rolled back if necessary.

2. **Versioning Across Environments:** Models typically go through multiple environments (development, staging, production) before final deployment. Each version of the model should be appropriately tagged to reflect the environment it is deployed in, ensuring consistency across all stages of development.

3. **Reverting to Previous Versions:** In case a newly deployed model introduces unexpected behavior or performance issues, version control allows for quick rollback to a previous stable version. This minimizes disruptions and mitigates risks associated with deploying updated models.

Some of the benefits of model versioning are as follows:

1. **Improved Traceability:** Teams can track exactly when, how, and why changes were made to the model, allowing for better auditing and compliance, especially in regulated industries like finance.

2. **Reproducibility:** Version control ensures that model performance can be reproduced and validated, even across different environments.

3. **Risk Mitigation:** If a new model version underperforms or introduces errors, versioning allows for a quick rollback to a previous, stable model, minimizing the impact on operations.

Caution Monitor Latency and Resource Consumption—LLMs can be computationally expensive. Ensure that new deployments do not introduce excessive latency or consume disproportionate amounts of memory and compute resources.

Automated Testing

The goal is to ensure that new or updated LLMs function as expected before they are deployed to production, reducing the risk of errors. Automated testing is a crucial step in the deployment pipeline, particularly for LLMs, where models can be complex, and errors in production can have significant consequences. Automated tests validate that the model meets predefined criteria for accuracy, performance, and compliance before deployment. In financial applications, where even small errors can lead to financial loss or compliance issues, automated testing is indispensable.

Note Testing Should Include Edge Cases—Automated testing should go beyond standard scenarios to include edge cases, adversarial inputs, and worst-case performance conditions to ensure robustness.

The following are the types of automated tests for LLMs:

1. **Unit Testing:** Focuses on individual components of the model, such as input/output functions, ensuring that each part operates correctly in isolation. For example, tests might verify that the LLM properly processes certain financial terms or market data inputs.

2. **Integration Testing:** Tests how different components of the LLM work together. This is especially important for models that interact with other systems (e.g., pulling financial data from external databases). Integration testing ensures that these interactions do not introduce errors or unexpected behaviors.

3. **Performance Testing:** Ensures that the LLM performs within acceptable limits, especially in real-time applications where latency and throughput are critical. Performance tests simulate live environments to assess how the model handles load, scalability, and resource usage.

4. **Compliance Testing:** In industries like finance, compliance with regulatory standards is critical. Automated tests can verify that the model adheres to these regulations, checking for biases, transparency, and auditability. Some of the benefits of automated testing are as follows:

    1. **Increased Reliability:** Automated tests ensure that the LLM functions correctly before deployment, reducing the likelihood of errors.

    2. **Faster Deployment:** Automated testing speeds up the validation process, allowing models to move through the pipeline quickly.

CHAPTER 4 DEPLOYMENT STRATEGIES FOR LLMS

3. **Compliance Assurance:** In industries with strict regulations, automated testing ensures that all compliance requirements are met before deployment, reducing the risk of fines or sanctions.

Tip Use Blue-Green or Canary Deployments—To ensure a smooth rollout, use blue-green or canary deployment strategies. These techniques allow you to test new models on a subset of traffic before full deployment, reducing the risk of system-wide failures.

Canary Deployments

The goal is to gradually roll out new models to production environments in a controlled manner, reducing the risk of widespread issues. Canary deployments are a strategy used to introduce new models into production environments incrementally. Rather than deploying the new LLM to the entire system at once, it is first deployed to a small subset of users or systems. This "canary" group allows the team to monitor the performance and behavior of the new model before rolling it out to the broader user base. Canary deployments work in the following manner:

1. **Small-Scale Rollout:** The new model is deployed to a small segment of the user base or to non-critical systems. During this phase, the performance of the model is closely monitored for any signs of issues, such as increased latency, unexpected outputs, or resource overuse.

2. **Performance Monitoring:** During the canary phase, detailed metrics are gathered to assess how the new model compares to the current production model. This includes monitoring for errors, compliance violations, and performance degradation.

3. **Full Rollout or Rollback:** If the canary deployment performs as expected, the model is gradually rolled out to the rest of the system. If issues are detected, the deployment can be paused, and the model can be rolled back to the previous version, ensuring that disruptions are minimized. Some of the benefits of canary deployments are the following:

1. **Reduced Risk:** Canary deployments reduce the risk of system-wide failures or performance degradation by limiting the exposure of the new model during its initial deployment phase.

2. **Real-Time Feedback:** Teams can gather real-time feedback on the model's performance in a production environment, allowing for adjustments or rollbacks if needed.

3. **Gradual Rollout:** The model can be rolled out in stages, giving teams the flexibility to monitor its impact and performance before fully committing to the deployment.

Caution Ensure Regulatory Compliance—Any automated deployment process should be aligned with financial regulations, ensuring that updates do not introduce biases or compliance risks.

A well-designed deployment pipeline is essential for ensuring that LLMs can transition seamlessly from development to production. By implementing robust CI/CD practices, maintaining proper version control, conducting automated testing, and utilizing canary deployments, organizations can deploy LLMs with confidence, knowing that the risks are minimized and performance is optimized.

CHAPTER 4 DEPLOYMENT STRATEGIES FOR LLMS

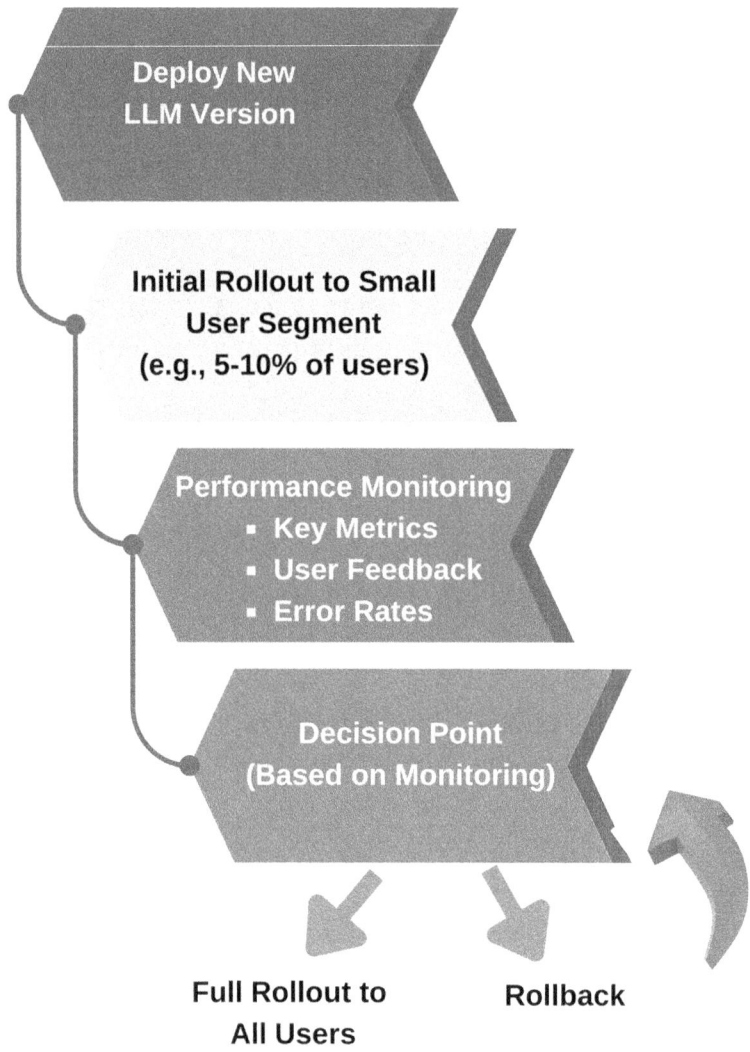

Figure 4-2. Canary Deployment Workflow

In Figure 4-2, the process begins with the Deploy New LLM Version, where a new iteration of the large language model is introduced into the deployment environment. Following this, the initial rollout to a small user segment involves deploying the new LLM to only a limited portion of the user base, which strategically minimizes potential disruptions or negative impacts. Performance monitoring is then crucial, encompassing several key aspects: Key metrics are meticulously tracked, including response time, accuracy, throughput, and resource utilization, to quantify the LLM's efficiency and effectiveness. User feedback is actively gathered from the small user segment to gain qualitative insights into their experience with the new version. Additionally, error rates are monitored to identify any failures, unexpected behavior, or increased occurrence of errors. At the decision point, a determination is made based on the comprehensive performance monitoring results. If the new LLM version performs satisfactorily and aligns with predefined criteria, a full rollout is initiated, making it available to all users. Conversely, if significant issues or regressions are observed, a rollback is executed, reverting the system to the previous, stable version. Finally, if the decision is to proceed, a full rollout to all users occurs, wherein the new LLM version is deployed across the entire user base, completing the deployment process. This carefully designed workflow enables a controlled and secure deployment of new LLM versions, effectively mitigating risks and ensuring a seamless transition for all users.

In the next section, we will explore the importance of monitoring and logging in maintaining the health and reliability of LLMs once they are deployed in production environments.

Note Model Drift Detection is Crucial—CI/CD alone is not enough; integrate model monitoring tools to detect drift, ensuring that deployed LLMs remain accurate and relevant.

Monitoring and Logging

Deploying large language models (LLMs) is only the first step toward building efficient and reliable systems. Once in production, continuous monitoring and logging are essential to ensure the long-term performance, stability, and reliability of the LLM. In industries like finance, where LLMs are used for critical tasks such as fraud detection, risk assessment, and customer service automation, real-time monitoring can help detect issues early and minimize downtime. In this chapter, we will explore the key aspects of monitoring and logging, focusing on performance metrics, error tracking, resource monitoring, and anomaly detection.

Note Monitoring Includes Infrastructure, Not Just Models—Track not only the model's performance but also infrastructure metrics such as CPU, memory, and GPU usage to prevent resource bottlenecks.

CHAPTER 4   DEPLOYMENT STRATEGIES FOR LLMS

Performance Metrics

The goal is to track and measure critical performance indicators to ensure the LLM is operating at optimal efficiency and adhering to industry standards, particularly in time-sensitive environments like finance.

Caution   Avoid Excessive Logging—While comprehensive logging is important, excessive logging can slow down the system and increase storage costs. Implement log rotation and retention policies to manage log size effectively.

Performance metrics provide a real-time view of how well the LLM is functioning in production. These metrics can highlight areas where the model might be underperforming, whether due to excessive latency, bottlenecks, or throughput issues. For financial institutions, performance metrics are not just about speed—they are also about meeting strict compliance and accuracy requirements. Some of the key performance metrics to monitor are the following:

1. **Response Time:** This metric measures how long it takes for the LLM to process a query and return a result. In high-frequency trading or real-time fraud detection, slow response times can lead to missed opportunities or increased risks. Tracking response times ensures that the LLM can handle real-time demands efficiently.

2. **Latency:** Latency refers to the delay between a request being sent and the model's response being received. Low latency is crucial for applications like customer service chatbots, where quick responses are expected. Monitoring latency helps in identifying potential slowdowns in the system.

3. **Throughput:** This metric measures the number of queries or requests the LLM can handle within a given time frame. Monitoring throughput helps identify whether the system is capable of scaling to meet high demand, which is common in financial systems with fluctuating workloads.

4. **Error Rate:** Error rate tracks the number of failed or incorrect responses generated by the LLM. High error rates can indicate underlying issues with the model's architecture, data handling, or integration with external systems. In finance, a high error rate could have severe consequences, including regulatory violations or financial loss.

Tip Use Centralized Logging Systems—Implement a centralized logging solution (e.g., ELK Stack, Google Cloud Logging, Datadog) to consolidate logs from multiple sources and enable efficient troubleshooting.

Monitoring performance metrics is essential in ensuring that LLMs in the finance industry remain compliant with strict regulatory standards, such as maintaining specific response times and error thresholds. These metrics provide critical insights that enable real-time decision-making, allowing organizations to quickly adjust resources or update models as needed. Additionally, continuous monitoring helps detect early signs of performance degradation, enabling teams to take proactive measures before the model's performance declines under real-time operational conditions. This ensures the system remains reliable, efficient, and responsive.

Caution Ensure Data Privacy in Logging—Avoid logging sensitive user data, especially in financial applications where compliance with regulations such as GDPR and PCI DSS is required. Use data masking and anonymization techniques when necessary.

Error Tracking

The goal is to identify, log, and address errors that occur in real-time, ensuring that the LLM continues to provide accurate and reliable results. In any deployment, errors are inevitable. However, in financial systems, errors can have far-reaching consequences, from incorrect loan approvals to missed fraud signals. Tracking and logging errors in real-time are crucial to maintaining the LLM's accuracy and ensuring its continued reliability. The different types of error tracking are the following:

1. **Unexpected Outputs:** Sometimes the LLM may generate outputs that are inconsistent with the expected results. For instance, in a financial report generation system, the LLM might misinterpret a data point or miscalculate a financial ratio. Monitoring these errors helps ensure the LLM is producing accurate results.

2. **Degraded Performance:** Even if the model generates correct outputs, degraded performance can still affect the overall system. For example, if the LLM slows down under a high load, or if resource constraints cause bottlenecks, error logs will capture these events for future analysis.

3. **System Crashes**: In rare cases, the LLM might crash due to insufficient resources, conflicts with other systems, or unhandled exceptions. Tracking and logging these crashes helps identify the root causes and allows for proactive troubleshooting.

4. **Compliance Violations:** In finance, errors that lead to regulatory non-compliance can have serious consequences, including fines and legal actions. Continuous error tracking ensures that any outputs that deviate from compliance standards are flagged and corrected.

Error tracking is crucial for maintaining the reliability and accuracy of LLMs, especially in industries like finance where precision is critical. By logging errors in real-time, teams can proactively address issues before they escalate into larger system-wide problems. Continuous error tracking helps ensure that the model maintains its accuracy, minimizing the risk of incorrect outputs. Additionally, detailed error logs provide essential data for root cause analysis, enabling teams to identify the underlying issues and prevent similar problems from occurring in the future, thus ensuring long-term system stability and performance.

Resource Monitoring

The goal is to optimize the use of system resources such as CPU, memory, and GPU, ensuring that the LLM operates efficiently without overloading the infrastructure. LLMs are resource-intensive, often requiring significant computational power to process large volumes of data and generate accurate predictions. Monitoring resource usage helps ensure that the system is operating efficiently, without wasting resources or straining the infrastructure. The key resources to monitor are the following:

1. **CPU Usage:** LLMs typically rely on multiple processors to handle their computations. Monitoring CPU usage ensures that the model is efficiently utilizing available processing power. High CPU usage may indicate that the system is under stress and might require scaling or optimization.

2. **Memory Usage:** LLMs can be memory-intensive, especially when handling large datasets or complex queries. Monitoring memory usage helps prevent the model from running out of memory, which can cause crashes or significant slowdowns.

3. **GPU Usage:** In many cases, LLMs rely on GPUs for faster computation, especially in real-time applications. Monitoring GPU usage ensures that the model is efficiently using its computational resources while also identifying any bottlenecks that might be limiting performance.

4. **Disk I/O:** Disk input/output (I/O) operations can become a bottleneck if the system relies heavily on data stored on disk. Monitoring disk I/O helps identify performance issues related to reading and writing large amounts of data.

Resource monitoring is essential for managing costs and maintaining system efficiency, particularly in cloud-based environments where computational power is billed based on usage. By continuously tracking resource utilization, organizations can prevent system overloads that could lead to slowdowns or crashes, ensuring that LLMs operate smoothly. Additionally, monitoring resource usage enables teams to make informed decisions about scaling the system or optimizing the model, ensuring that performance remains high while keeping operational costs under control.

Tip Monitor Model Drift Regularly—Establish automated checks to detect data drift, ensuring that your LLM remains aligned with current market and financial trends.

Anomaly Detection

The goal is to set up real-time systems that automatically detect anomalies in model performance, behavior, or resource usage, allowing for quick identification and resolution of potential issues. Anomalies are unexpected behaviors or outputs that deviate from normal patterns. In financial systems, detecting anomalies quickly is critical to prevent incorrect predictions, performance degradation, or even compliance violations. Anomaly detection systems provide early warnings, allowing teams to intervene before problems escalate. The key areas for anomaly detection are the following:

1. **Performance Anomalies:** Sudden spikes in response times or increased error rates are clear indicators of performance degradation. Real-time anomaly detection systems can flag these deviations for immediate investigation.

2. **Data Drift:** In financial applications, the underlying data often changes over time. Data drift occurs when the distribution of input data shifts significantly from what the model was trained on. This can lead to poor predictions and unreliable outputs. Anomaly detection systems can identify data drift and trigger model retraining.

3. **Resource Anomalies:** Sudden increases in CPU, memory, or GPU usage can indicate that the model is struggling to handle the current workload. Monitoring systems can flag these spikes as anomalies, helping teams to adjust resources or optimize the model.

4. **Compliance Anomalies:** In finance, it is essential to ensure that the LLM adheres to regulatory requirements. Anomaly detection systems can flag any outputs that violate compliance rules, allowing teams to correct issues before they lead to fines or legal consequences.

Note Anomaly Detection Should Be Adaptive—Implement machine learning-based anomaly detection to dynamically adjust thresholds based on historical patterns and real-time conditions.

Anomaly detection plays a crucial role in preventing system failures by identifying irregularities early, allowing teams to address minor issues before they escalate into larger problems that could impact the entire system. It also ensures consistent performance, as anomalies often signal performance degradation, which can compromise the reliability of LLMs. In the fast-paced finance industry, where data and market conditions can shift rapidly, detecting anomalies such as data drift enables teams to retrain models promptly, ensuring they remain accurate, reliable, and aligned with current market dynamics.

Continuous monitoring and logging are critical components of deploying and maintaining LLMs, particularly in high-stakes industries like finance. By tracking performance metrics, logging errors, monitoring resource usage, and implementing real-time anomaly detection, organizations can ensure that their LLMs operate reliably, efficiently, and in compliance with industry standards. Monitoring and logging not only help identify and resolve issues early but also provide a framework for continuous improvement. In the next section, we will explore performance optimization strategies to ensure that LLMs remain efficient and scalable in production environments.

Caution   Beware of Alert Fatigue—Too many alerts can overwhelm your monitoring team, causing critical warnings to be overlooked. Use smart alerting techniques such as rate-limiting and prioritization.

Performance Optimization

Deploying large language models (LLMs) in a high-demand environment like finance presents unique performance challenges. Financial systems require models that are not only accurate but also highly efficient, scalable, and responsive in real time. Performance optimization is critical for ensuring that LLMs can handle large volumes of data, provide quick responses, and operate within financial industry constraints such as compliance and cost management. This section explores key strategies for optimizing the performance of LLMs in production, with a focus on scaling, model compression, latency reduction, and workload balancing.

CHAPTER 4 DEPLOYMENT STRATEGIES FOR LLMS

Scaling LLMs

The goal is to enable LLMs to handle increasing volumes of data and requests while maintaining performance and reliability. Scaling is one of the most important aspects of performance optimization, especially for LLMs deployed in financial systems. As data volumes and user demands grow, LLMs must be able to scale efficiently without sacrificing response times or accuracy. There are two primary scaling strategies: vertical scaling and horizontal scaling.

Vertical Scaling (Scaling Up):

Vertical scaling offers the advantage of being relatively easy to implement, as it requires minimal changes to the existing system architecture. By adding more resources, such as CPU or memory, to a single machine, performance can be enhanced without the need for complex reconfigurations. However, this approach has its limitations, as there is only so much hardware that can be added to a single machine before hitting diminishing returns. Additionally, vertical scaling can become expensive, particularly in cloud environments where costs are closely tied to resource consumption, making it less sustainable for long-term scalability needs.

Tip Implement Hybrid Scaling Strategies—Use a mix of vertical and horizontal scaling to balance performance and cost. Vertical scaling is useful for improving single-instance efficiency, while horizontal scaling ensures high availability and resilience.

Horizontal Scaling (Scaling Out):

Horizontal scaling involves distributing the workload across multiple machines or servers. By using more machines, the system can handle a much larger volume of requests and process data in parallel. Horizontal scaling offers greater flexibility by allowing systems to scale indefinitely through the addition of more machines, making it ideal for distributed systems like cloud environments where multiple instances of the LLM can run in parallel. This approach enables organizations to handle increasing workloads without the limitations of a single machine. However, implementing horizontal scaling can be complex, as it requires proper load balancing and coordination between the various instances of the model. Additionally, ensuring data consistency and synchronization across multiple nodes becomes a critical challenge, adding to the complexity of the system.

CHAPTER 4   DEPLOYMENT STRATEGIES FOR LLMS

Best Practices for Scaling

Effective scaling requires implementing best practices such as load balancing and auto-scaling to optimize performance and ensure system stability. Load balancing is crucial for distributing requests evenly across all instances of the LLM, preventing any single machine from becoming overwhelmed and ensuring optimal resource utilization. By routing traffic based on system load and availability, load balancers help maintain consistent performance. Auto-scaling mechanisms further enhance scalability by dynamically adjusting the number of machines in response to real-time demand. This is particularly valuable in financial systems, where workloads can fluctuate significantly during peak times, such as market openings or closings, ensuring that the system can handle sudden spikes without compromising performance.

Model Compression and Quantization

The goal is to reduce the size of LLMs for more efficient use of resources without sacrificing accuracy. LLMs are often large and resource-intensive, making them expensive to run, especially in cloud environments. Model compression and quantization techniques can significantly reduce the size of models, improving both speed and cost-efficiency while maintaining high levels of accuracy.

Tip   Leverage Model Compression for Cost Savings—Applying techniques like pruning, quantization, and model distillation can significantly reduce inference costs in cloud environments while maintaining near-original accuracy.

Model Compression: Model compression involves techniques that reduce the number of parameters or the size of the neural network. This can be done without significantly impacting the performance of the LLM, allowing it to process data faster and use fewer computational resources.

As shown in Table 4-2, model compression methods like pruning, distillation, and quantization help reduce size and improve efficiency. Each method offers trade-offs, balancing accuracy, speed, and resource needs.

Table 4-2. Model Compression and Quantization Methods

Method	Description	Trade-offs
Pruning	Removing non-critical neurons.	Reduces size but might slightly affect accuracy.
Distillation	Training a smaller model to mimic a larger one.	Maintains accuracy but requires additional training time.
Quantization	Reducing the precision of numbers used in the model.	Improves speed but may lead to slight performance degradation

Pruning: This technique involves removing parts of the model (such as neurons or connections) that contribute little to the final output. Pruning reduces the overall size of the model and can lead to faster inference times.

Distillation: Model distillation is a technique where a smaller model (the "student") is trained to mimic the behavior of a larger, more complex model (the "teacher"). The smaller model is often faster and less resource-intensive but can still achieve near-identical performance to the original.

Caution Beware of Over-Compression—Excessive quantization or pruning can lead to loss of accuracy, especially in compliance-critical tasks. Always validate model performance post-compression.

Quantization

Quantization is a technique that reduces the precision of numbers used to represent a model's parameters, allowing for faster computation and lower memory usage. Typically, deep learning models use 32-bit floating-point numbers, but quantization reduces this precision to 16-bit or even 8-bit integers, which are quicker to process. This approach offers significant benefits, such as reducing both the model size and computational costs, while often maintaining performance with minimal loss in accuracy. However, challenges can arise, as some models may experience accuracy degradation after quantization. Therefore, careful fine-tuning, testing, and evaluation are crucial to ensure the compressed model performs effectively.

Best Practices for Model Compression and Quantization:

To ensure optimal performance when using model compression and quantization, certain best practices should be followed. Fine-tuning after compression is essential, as it helps restore any accuracy that may have been lost during the compression process by retraining the model on the target dataset. Additionally, employing hybrid approaches—such as combining pruning with quantization—can deliver more effective results by improving both the compression rate and model accuracy, ensuring that the model remains efficient without sacrificing performance.

Latency Reduction

The goal is to minimize delays in real-time applications by optimizing the model, hardware, and network infrastructure. In financial systems, where decisions must be made in real time (e.g., trading algorithms, fraud detection, customer service), reducing latency is critical. Even minor delays can lead to significant financial losses or missed opportunities. Latency can occur at multiple stages: within the model itself, at the hardware level, or due to network constraints.

Hardware Optimizations:

> **GPUs and TPUs:** Using GPUs (Graphics Processing Units) or TPUs (Tensor Processing Units) can accelerate the computation required by LLMs, particularly for parallel processing tasks. Hardware accelerators like TPUs are designed specifically for AI workloads and can offer significant performance improvements over traditional CPUs.
>
> **Optimized Libraries:** Leveraging optimized libraries like CUDA (for NVIDIA GPUs) or CuDNN can further improve the performance of LLMs, reducing inference time and speeding up model execution.

Software Optimizations:

> **Batch Processing:** Instead of processing one query at a time, LLMs can be optimized to handle multiple queries simultaneously through batch processing. This approach increases throughput and reduces the overall response time per query.

CHAPTER 4 DEPLOYMENT STRATEGIES FOR LLMS

Efficient Algorithms: Optimizing the underlying algorithms used for inference can also reduce latency. For example, beam search or greedy search algorithms can be optimized to provide faster responses during text generation tasks.

Network Optimizations:

Content Delivery Networks (CDNs): Using CDNs to cache and serve model outputs can reduce the time it takes for results to be delivered to end users, especially in globally distributed environments.

Network Caching: Caching frequently requested queries or responses at the network level reduces the need to recompute results, significantly lowering response times for recurring queries.

To reduce latency effectively, implementing best practices is essential. Profiling and bottleneck analysis involve regularly analyzing the system to pinpoint the sources of latency, whether they arise from the model, hardware, or network infrastructure. Once identified, targeted optimizations can be applied to address these specific issues. Additionally, using edge computing for ultra-low latency applications can significantly reduce data travel time by deploying LLMs closer to the end user, minimizing delays in communication between the server and the user.

Additionally, using edge computing for ultra-low latency applications can significantly reduce data travel time by deploying LLMs closer to the end user, minimizing delays in communication between the server and the user. This can be implemented using edge deployment solutions such as **AWS Greengrass** or **Google Cloud Edge TPU**, which allow models to run inference locally on edge devices, enabling real-time responsiveness in critical applications.

Table 4-3 shows a comparison of commonly used edge deployment platforms that are well-suited for running LLMs and AI workloads at the edge.

Table 4-3. Edge Deployment Options for Ultra-Low Latency LLM Inference

Platform	Description	Ideal Use Cases	Strengths
AWS Greengrass	Run LLM components on local edge devices while syncing with the AWS cloud.	IoT, industrial monitoring, smart buildings	Integrates with AWS services, supports local Lambda functions
GCP Edge TPU	Specialized ASIC hardware for running TensorFlow Lite models at the edge.	Smart cameras, anomaly detection, offline NLP	High speed, energy efficient, optimized for Coral devices
NVIDIA Jetson	AI edge computing platform with GPU acceleration for heavier models.	Robotics, autonomous machines, edge LLM tasks	Supports PyTorch/ONNX, powerful onboard GPU compute

Caching and Load Balancing

The goal is to optimize the distribution of workloads and reduce computational redundancy by caching frequently requested outputs. Caching and load balancing are essential techniques to ensure that LLMs operate efficiently in production environments. These strategies not only improve performance but also reduce the overall computational load on the system, making it easier to scale while maintaining low latency and high availability.

> **Tip** Use Adaptive Load Balancing—Instead of static load balancing methods, use real-time adaptive load balancers that distribute traffic dynamically based on model performance, CPU/GPU utilization, and request priority.

Caching involves storing frequently requested outputs so that they can be reused without recalculating the result. In financial applications, certain queries may be repeated often, such as retrieving market reports or generating specific financial summaries. By caching these responses, the LLM can return results almost instantaneously for repeated queries, saving both time and computational resources.

In-Memory Caching: Using in-memory storage systems (e.g., Redis or Memcached) allows for fast retrieval of cached data, improving response times for frequently accessed outputs.

Database Caching: For more complex queries that require database access, caching the results of previous database queries can reduce the need for repeated access to the database, speeding up the overall process.

Load balancing distributes incoming requests across multiple servers or instances of the LLM. This ensures that no single instance is overwhelmed, which can lead to slowdowns or crashes, particularly during periods of high demand.

As shown in Table 4-4, caching and load balancing approaches such as in-memory caching, round-robin, and dynamic load balancing help improve efficiency and reliability. These techniques ensure faster access, fair distribution, and optimal performance in financial systems.

Table 4-4. Caching and Load Balancing Approaches

Approach	Description	Example
In-Memory Caching	Stores frequently accessed queries in memory.	Market reports retrieval for trading desks.
Round-Robin Load Balancing	Distributes requests evenly across all servers.	Ensures fair workload distribution in fraud detection systems.
Dynamic Load Balancing	Adjusts traffic routing based on real-time server load.	Helps maintain optimal performance in high-traffic financial systems.

Round-Robin Load Balancing: A simple load balancing method where requests are distributed evenly across available servers in a rotating fashion.

Dynamic Load Balancing: More advanced load balancers can dynamically adjust the distribution of requests based on real-time server loads, routing traffic to the least busy servers for optimal performance.

Best Practices for Caching and Load Balancing:

Cache Expiration Policies: Implement expiration policies to ensure that cached data remains up-to-date, particularly in environments where data changes frequently, such as in finance.

Health Monitoring: Regularly monitor the health and performance of each server in the load balancing pool to ensure that traffic is being distributed evenly and no servers are underperforming.

Performance optimization is a critical aspect of deploying LLMs in production, particularly in resource-intensive environments like finance. By implementing strategies for scaling, model compression, latency reduction, and workload balancing, organizations can ensure that their LLMs operate efficiently, deliver fast responses, and remain cost-effective. These techniques help maintain the reliability and performance of LLMs, enabling them to handle the growing demands of real-time financial applications.

In the next section, we will explore the LLM Ops Lifecycle, a systematic framework that integrates all the key strategies discussed—deployment pipelines, monitoring and logging, and performance optimization—into a cohesive approach for managing LLMs in production environments. This lifecycle approach is particularly critical for the finance industry, where continuous model improvement, regulatory compliance, and real-time system reliability are essential. By understanding and applying the stages of this lifecycle, organizations can ensure that their LLMs operate efficiently, adapt to evolving data and regulations, and consistently deliver value in high-demand financial systems.

Tip Automate Model Revalidation and Retraining—Set up scheduled retraining pipelines to ensure LLMs stay relevant to financial market shifts and regulatory updates.

Understanding the Lifecycle

The implementation and operation of LLMs in finance bring unique opportunities and challenges. This section describes the LLM Operations (LLM Ops) lifecycle, a systematic approach to the development, deployment, and continuous monitoring of LLMs deployed for financial applications.

CHAPTER 4 DEPLOYMENT STRATEGIES FOR LLMS

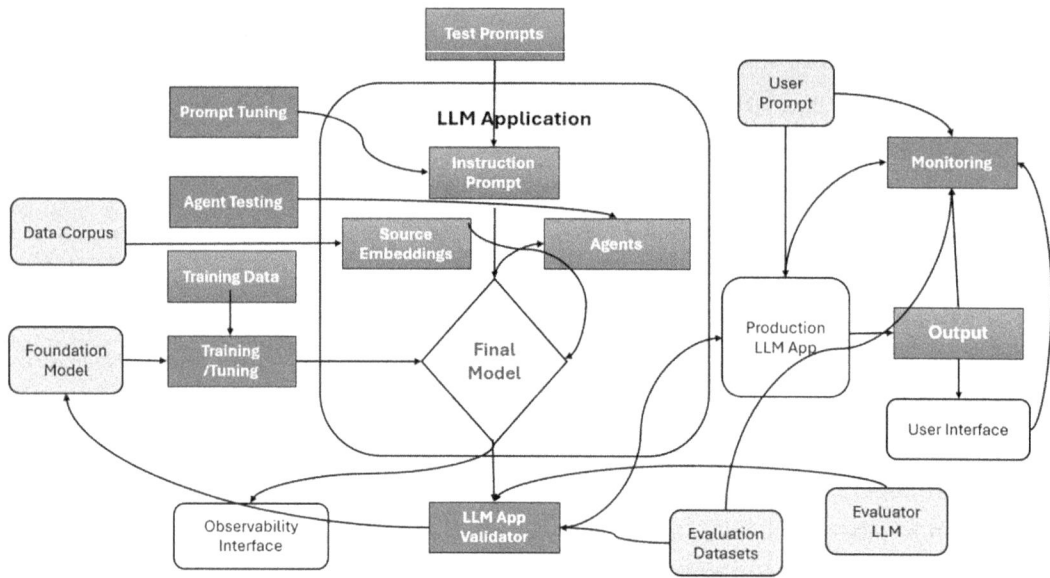

Figure 4-3. Deployment Lifecycle of LLMs

Figure 4-3 represents the LLM Ops lifecycle, which outlines the key stages and interactions between model training, validation, and deployment. While the general flow of LLM Ops is consistent across industries, applying these operations in finance requires special attention to regulatory compliance, security, data sensitivity, and ethical considerations. Let's explore the critical components of this lifecycle and how they specifically cater to the needs of financial applications.

Foundation Model

At the heart of the LLM Ops lifecycle is the foundation model. These models serve as the backbone of LLM-based applications, capable of generating insights, summarizing large volumes of data, and even forecasting financial trends.

> **Data Corpus:** LLMs for finance require an extensive corpus of domain-specific data, including financial reports, market trends, investment research, and regulatory filings. This ensures the model is proficient in understanding financial terminology and industry jargon.

Training Data: Training financial LLMs involves curated datasets, such as historical market data, pricing trends, and even textual data from earnings calls or analyst reports. Careful preprocessing is crucial to avoid overfitting and to maintain generalizability across financial contexts.

Training/Tuning: Fine-tuning the foundation model involves adjusting the model parameters to focus on specific financial use cases, such as risk assessment, fraud detection, or investment recommendations. The process includes supervised and unsupervised techniques, using financial experts' input to refine the model.

LLM Application

The LLM application layer is where various components come together to create an operable system that can interact with user prompts and provide actionable insights.

Source Embeddings: Financial LLMs require embeddings that reflect the nuances of the domain, such as stock ticker symbols, macroeconomic indicators, and specific corporate actions. Embeddings condense these financial data into vectors, which the model can interpret to generate meaningful outputs.

Instruction Prompts: A critical element in LLMs is the prompt system. For financial models, instruction prompts guide the model to perform specific tasks such as generating earnings summaries, answering regulatory compliance questions, or providing risk assessments.

Agents: These are automated systems that handle interactions with external systems, such as querying databases for the latest financial data or triggering alarms based on market volatility predictions. For financial institutions, agents may serve roles in managing portfolio allocations or detecting fraudulent activities.

Prompt Tuning and Agent Testing: These steps ensure that the financial LLM operates within predefined boundaries, producing accurate and reliable results. For instance, in a banking context, it is essential that prompt tuning leads to accurate loan approval predictions or accurate identification of high-risk transactions.

Caution   Monitor for Data Drift Continuously—Financial models must stay aligned with market and regulatory changes. Regular drift detection prevents performance degradation due to outdated training data.

Validation and Evaluation

Before deploying an LLM in the financial domain, extensive LLM app validation is required. This step involves running rigorous tests on the model to ensure that its predictions or suggestions align with financial regulations and industry standards.

> **Evaluation Datasets:** These datasets help to assess the model's accuracy and performance. In finance, this might include testing the model on previously unseen market data, regulatory updates, or extreme market scenarios (e.g., during financial crises). It is also advisable to include sandbox datasets, curated environments that simulate rare or high-risk edge cases, to stress-test the model under extreme or atypical conditions.
>
> **Evaluator LLM:** An evaluator LLM can be used to test the output from the primary LLM. This ensures that financial models do not produce biased outputs or misleading information, which is crucial for maintaining compliance with regulations such as the SEC's guidelines or international banking standards.

Production and Monitoring

Once an LLM has passed validation, it can be deployed as a production LLM app. In finance, where accuracy and regulatory compliance are paramount, constant monitoring and feedback mechanisms are necessary.

Monitoring: The production model is closely monitored for anomalies, such as data drift, which could result in inaccurate forecasts or risk assessments. In finance, this is vital to ensure that the model is not producing skewed results due to changes in market behavior or external economic shocks.

Human Feedback: Financial analysts or regulatory officers often provide real-time feedback to the model's outputs. This feedback loop ensures that the model's decisions align with the evolving nature of financial regulations and market conditions.

User Interface (UI): Financial applications require intuitive user interfaces that provide clear and actionable insights. A UI might present key risk indicators, portfolio analysis, or investment suggestions, enabling financial experts to make informed decisions swiftly.

Caution Monitor for Data Drift Continuously—Financial models must stay aligned with market and regulatory changes. Regular drift detection prevents performance degradation due to outdated training data.

Iterative Improvement and Fail-Safe Mechanisms

Financial LLMs must be capable of evolving with new data, regulations, and market dynamics. The LLM Ops lifecycle includes mechanisms to ensure continuous improvement and fail-safes to catch potential errors before they escalate into critical failures.

Observability Interface: This interface captures the model's real-time performance metrics and errors, alerting engineers or financial analysts to potential issues in the production model. For example, if the model shows a decline in accuracy during market shocks, adjustments can be made rapidly.

Human Feedback Loop: In financial operations, human oversight is indispensable. Continuous feedback from finance professionals helps tune the model further and adjust for any biases or inaccuracies, particularly in complex tasks like regulatory compliance or market forecasting.

> **Revalidation and Retraining:** If any issues arise or significant changes occur in the financial data environment, the LLM can be retrained with updated data or improved evaluation metrics. This ensures that the model maintains its efficacy and accuracy over time.

Implementing LLM Ops in finance is not just about deploying models; it's about ensuring continuous alignment with financial regulations, market shifts, and client needs. By understanding and leveraging the LLM Ops Lifecycle, financial institutions can build robust systems that improve decision-making, enhance operational efficiency, and reduce risk.

In the following chapters, we will explore case studies of financial institutions that have successfully implemented LLM Ops, highlighting the challenges they faced and the outcomes they achieved.

Conclusion

In this chapter, we have explored the critical strategies for deploying large language models (LLMs) in production environments, with a particular focus on the finance industry. The deployment of LLMs goes beyond simply implementing well-trained models; it requires a robust and comprehensive approach that encompasses efficient deployment pipelines, continuous monitoring and logging, and performance optimization. By implementing CI/CD practices, model versioning, automated testing, and canary deployments, organizations can ensure a smooth transition from development to production, minimizing downtime and risk.

The importance of monitoring and logging cannot be overstated, as these practices ensure real-time performance tracking, error detection, and resource optimization, enabling models to operate reliably and meet regulatory requirements. Performance optimization, through strategies like scaling, latency reduction, model compression, and workload balancing, ensures that LLMs remain efficient and scalable even under high demand, particularly in real-time financial applications.

Additionally, we introduced the LLM Ops Lifecycle, which integrates all these strategies into a cohesive framework for the ongoing development, deployment, and monitoring of LLMs. This lifecycle approach is essential in finance, where models must continuously evolve to meet regulatory compliance, adapt to new data, and maintain accuracy in dynamic environments.

By applying the principles and strategies outlined in this chapter, organizations can ensure their LLMs not only operate effectively but also remain flexible and compliant in the fast-paced world of finance. In the following chapters, we will further explore case studies of financial institutions that have successfully implemented these deployment strategies, offering valuable insights and real-world lessons on overcoming challenges and achieving operational excellence.

Points to Remember

- **CI/CD Automation:** Implement continuous integration and deployment (CI/CD) to ensure seamless updates and minimal downtime when deploying LLMs.

- **Model Versioning:** Keep track of multiple model versions to ensure stability, transparency, and easy rollback if issues arise in production.

- **Automated Testing:** Use automated testing to validate model performance, accuracy, and compliance before deployment, reducing the risk of errors.

- **Canary Deployments:** Gradually roll out new models using canary deployments to monitor performance and mitigate risks before full deployment.

- **Real-Time Monitoring:** Continuously monitor key performance metrics like response time, latency, and error rates to maintain system health and compliance.

- **Error Tracking:** Implement real-time error logging to quickly address issues and ensure model accuracy in critical financial applications.

- **Resource Monitoring:** Track resource usage (CPU, memory, GPU) to prevent system overloads and optimize the performance of LLMs.

- **LLM Ops Lifecycle: Anomaly Detection:** Set up systems to detect anomalies such as data drift, ensuring timely interventions to prevent performance degradation.

- **Performance Optimization:** Use strategies like vertical/horizontal scaling, model compression, and latency reduction to enhance system efficiency and responsiveness.

CHAPTER 5

Ensuring Data Privacy and Security

Data privacy and security are fundamental to deploying large language models (LLMs) in the finance sector, where protecting sensitive information is paramount. This chapter provides a deep dive into the critical aspects of data privacy and security for LLM operations, emphasizing strategies to safeguard data throughout the LLM lifecycle. It begins by discussing data anonymization techniques such as masking and encryption, which are essential for protecting personally identifiable information (PII) and maintaining compliance with data protection laws.

Next, the chapter explores secure data storage solutions, including encryption methods and access control mechanisms, to prevent unauthorized access and ensure the integrity of financial data. Compliance with financial regulations, such as GDPR and CCPA, is also covered, along with a detailed overview of legal considerations when using AI in finance. By implementing these privacy and security measures, readers will be able to create a secure environment for LLM operations that meets both internal security standards and regulatory requirements.

Structure

This chapter covers the following topics:

- **Introduction to Data Privacy and Security:** Importance of privacy in financial LLMs, risks of data breaches, and compliance violations. Overview of anonymization, secure storage, and regulatory adherence.

- **Data Anonymization Techniques:** Methods like masking, generalization, suppression, and differential privacy. Applications in fraud detection and balancing privacy with usability.

- **Secure Data Storage:** Best practices for encryption (AES, RSA), access controls, multi-factor authentication, and secure cloud storage.

- **Compliance with Financial Regulations:** Key regulations (GDPR, CCPA), aligning LLM operations with compliance, data minimization, audits, and real-world examples.

- **Challenges and Best Practices**: Privacy vs. data utility, emerging threats (adversarial attacks), and best practices like protocol updates, compliance collaboration, and stakeholder education.

- **Future Trends in Privacy and Security:** Innovations like homomorphic encryption, federated learning, and evolving regulatory frameworks for financial data security.

Objectives

Data privacy and security are fundamental when deploying large language models (LLMs) in financial applications, where protecting sensitive customer and transactional data is a regulatory and operational priority. In this chapter, you will explore essential strategies to safeguard financial data while ensuring compliance with industry regulations such as GDPR, CCPA, and financial sector-specific standards. You will learn how to implement data anonymization techniques, including masking, generalization, suppression, and differential privacy, to protect personally identifiable information (PII) while maintaining data usability for LLM workflows.

This chapter will also guide you through best practices for secure data storage, covering encryption methods (AES, RSA), role-based access control, and multi-factor authentication to prevent unauthorized access and potential breaches. Additionally, you will gain insights into aligning LLM operations with compliance requirements, focusing on data minimization, audit trails, and transparent AI governance. The chapter will address challenges in balancing security with AI performance, highlighting threats like adversarial attacks and data leakage risks.

Looking ahead, you will also explore emerging trends in financial data security, such as homomorphic encryption and federated learning, and how these innovations can enhance LLM privacy-first workflows. By the end of this chapter, you will be equipped with the knowledge and tools to build secure, compliant, and trustworthy LLM-powered financial systems that protect both organizational integrity and customer trust.

Introduction

The exponential growth of artificial intelligence and the adoption of large language models (LLMs) in the financial sector have introduced groundbreaking opportunities to improve operational efficiency, decision-making, and customer engagement. However, these advancements come with a critical responsibility: ensuring data privacy and security. Financial systems handle highly sensitive information, including customer identities, transaction histories, and regulatory records, making data protection a top priority in LLM operations.

In the context of LLMs, data privacy and security are essential for maintaining trust, compliance, and operational effectiveness.

- **Protecting Sensitive Information:** Financial data includes personally identifiable information (PII), payment details, and proprietary business insights. Breaches of this data can result in severe financial and reputational damage.

    Example: If an LLM used for customer support inadvertently exposes sensitive account details, it could lead to identity theft or fraud.

- **Ensuring Compliance with Regulations:** Financial institutions are subject to strict data protection laws like GDPR, CCPA, and PCI DSS. LLM operations must align with these regulations to avoid penalties and maintain customer trust.

    Example: Compliance with GDPR's "right to be forgotten" requires financial institutions to delete personal data upon user request, even if it was used to train an LLM.

- **Maintaining Customer Trust:** Customers expect their data to be handled securely and responsibly. Ensuring privacy is a foundational step in building and maintaining trust.

Example: A bank leveraging LLMs for personalized financial advice must reassure customers that their financial data will not be misused or exposed.

- **Operational Continuity and Resilience:** Data security breaches can disrupt operations, cause downtime, and lead to significant recovery costs. By securing data, organizations ensure uninterrupted service delivery.

Example: A ransomware attack targeting a financial institution's LLM training pipeline could halt fraud detection systems, exposing the organization to substantial risk.

Integrating LLMs into financial workflows presents unique risks and challenges that must be mitigated effectively.

1. **Data Breaches and Unauthorized Access:**
   - LLMs rely on vast datasets, often stored in cloud environments. Without proper encryption and access controls, these datasets are vulnerable to breaches.
   - **Example**: An unsecured API endpoint for accessing LLM training data could expose sensitive transaction records to unauthorized parties.

2. **Model Inversion Attacks:**
   - Attackers can exploit LLMs to extract sensitive information from the model's training data.
   - **Example**: By querying an LLM, an attacker may reconstruct sensitive details, such as credit card numbers or account balances, used during training.

3. **Data Anonymization Challenges:**
   - Effective anonymization ensures data privacy but can reduce the utility of data for model training. Striking a balance between privacy and usability is a significant challenge.
   - **Example**: Excessive generalization in customer data anonymization may degrade an LLM's ability to generate accurate financial recommendations.

CHAPTER 5   ENSURING DATA PRIVACY AND SECURITY

4. **Regulatory Compliance Complexity**:

   - Financial institutions often operate across multiple jurisdictions, each with its own data protection laws. Aligning LLM operations with these diverse regulations is resource-intensive.

   - **Example:** An institution operating in both the EU and the US must comply with GDPR's stringent privacy standards while addressing CCPA's consumer rights requirements.

5. **Emerging Threats**:

   - Cyber threats evolve continuously, targeting data pipelines, storage systems, and model inference processes.

   - **Example**: Adversarial attacks may inject malicious data into an LLM's training pipeline, compromising model integrity.

As shown in Table 5-1, anonymization techniques such as masking, generalization, suppression, differential privacy, and pseudonymization protect sensitive data. Each method balances privacy with maintaining useful data utility in finance.

Table 5-1. Techniques for Anonymization

Technique	Description	Example in Finance	Impact on Data Utility
Masking	Replaces sensitive data with symbols or hashes.	Masking account numbers as XXXX-XXXX-XXXX-1234.	Retains format but hides real values.
Generalization	Replaces specific data with broader categories.	Age 28 is generalized to 25-35.	Reduces granularity but preserves trends.
Suppression	Removes sensitive data fields entirely.	Removing phone numbers from datasets.	Can remove important context.
Differential Privacy	Introduces statistical noise to protect individual data.	Adding slight variations to transaction amounts.	Preserves privacy while keeping useful patterns.
Pseudonymization	Replaces identifiers with random or reversible values.	Converting John Doe to User12345.	Enables tracking but prevents identification.

CHAPTER 5   ENSURING DATA PRIVACY AND SECURITY

This chapter provides a comprehensive framework for addressing the critical aspects of data privacy and security in LLM operations, with specific emphasis on the financial sector. Key areas of focus include

1. **Data Anonymization Techniques:** Practical methods to anonymize sensitive data while preserving its utility for training LLMs. Readers will learn about data masking, generalization, and differential privacy as strategies to protect customer information.

2. **Secure Data Storage:** Strategies for encrypting, partitioning, and managing data to ensure its security during storage and access. Topics include best practices for cloud-based storage and role-based access controls.

3. **Compliance with Financial Regulations:** Guidelines for aligning LLM operations with legal and regulatory requirements, such as GDPR and CCPA. This section highlights the importance of regular audits, data minimization, and maintaining documentation to demonstrate compliance.

As LLMs become integral to financial systems, prioritizing data privacy and security is no longer optional; it is a fundamental requirement. By addressing the risks and implementing the strategies outlined in this chapter, organizations can safeguard sensitive information, ensure compliance, and build resilient systems. This chapter provides actionable insights to help organizations navigate the complexities of securing LLM workflows, setting a foundation for ethical and sustainable AI operations in finance.

Data Anonymization Techniques

Data anonymization is a foundational strategy for protecting sensitive information in LLM workflows, particularly in the highly regulated financial sector. It involves transforming data in a way that ensures privacy while retaining its utility for tasks like model training and inference. This section explores the definition, common techniques, applications in LLMs, and challenges associated with data anonymization.

Data anonymization is the process of removing or obfuscating personally identifiable information (PII) and other sensitive data elements from datasets. The goal is to make it impossible (or highly unlikely) to identify individuals while preserving the dataset's usefulness for analytics, machine learning, and operational purposes. It is a foundational strategy for protecting sensitive information in LLM workflows, particularly in the highly regulated financial sector. It involves transforming data in a way that ensures privacy while retaining its utility for tasks like model training and inference. Key anonymization methods include **generalization**, **suppression**, and **k-anonymity,** a widely used privacy metric that ensures each individual is indistinguishable from at least $k - 1$ others within a dataset. This section explores the definition, common techniques, applications in LLMs, and challenges associated with data anonymization in production environments.

Additionally, **synthetic data generation** is emerging as a powerful privacy-preserving approach, enabling the creation of large-scale, statistically representative data without exposing real user information. This section explores the definition, common techniques, applications in LLMs, and challenges associated with data anonymization.

Significance in LLM Operations:

- **Privacy Protection:** Safeguards sensitive customer information from breaches or misuse.

- **Regulatory Compliance:** Ensures adherence to data protection laws like GDPR and CCPA, which require strict privacy measures.

- **Preserving Data Utility:** Allows organizations to use anonymized data for LLM training and operations without compromising security or violating privacy standards.

Example:

In a financial context, anonymizing transaction data may involve removing account numbers and replacing customer names with pseudonyms, enabling LLMs to analyze patterns without exposing sensitive details. Some of the common techniques for data anonymization are:

1. **Data Masking:** Replacing sensitive data with obfuscated values, such as using asterisks, hashes, or pseudonyms.

    - **Use Case:** Ideal for situations where data format must remain intact but actual values can be hidden.

    - **Example in Finance:** Masking credit card numbers as XXXX-XXXX-XXXX-1234 while retaining the last four digits for identification.

2. **Generalization and Suppression:** Generalization is replacing specific data points with broader categories to reduce identifiability. Suppression is removing sensitive data fields entirely.

    - **Use Case:** Useful when granular details are not essential for analysis.

    - **Example in Finance:**

        - Generalizing customer ages into ranges (e.g., "25-35" instead of "28").

        - Suppressing unnecessary data fields, such as phone numbers, from transaction records.

3. **Differential Privacy:** It is adding statistical noise to datasets to obscure individual data points while preserving overall patterns.

    - **Use Case:** Ensures robust privacy protection, especially for datasets used in large-scale LLM training.

    - **Example in Finance:** Introducing slight variations in transaction amounts to anonymize data while retaining patterns for fraud detection models.

    - **Advantages:** Allows organizations to share anonymized datasets without compromising privacy.

The following are the applications in LLM where these techniques are used:

1. **Anonymizing Training Data:** LLMs often require large datasets for training, which may include sensitive financial information. Anonymizing this data ensures compliance with privacy regulations while enabling effective model training.

    - **Example:** Preparing a dataset of transaction records for training a fraud detection LLM involves masking account details and generalizing geographic data to city-level granularity.

2. **Compliance with Privacy Laws:** Regulations like GDPR mandate anonymization when processing personal data for analytics or training purposes. Adopting techniques like data masking and differential privacy ensures that LLM operations align with these requirements.

    - **Example:** A compliance chatbot trained on anonymized regulatory documents ensures sensitive organizational details are not exposed.

3. **Collaboration Across Entities:** Anonymized data facilitates secure sharing of datasets between institutions for joint LLM training projects, enabling collaborative innovation without privacy risks.

    - **Example:** Banks sharing anonymized transaction data to improve a shared anti-money laundering (AML) model.

Some of the challenges are as follows:

1. **Balancing Data Utility and Privacy:** Excessive anonymization can degrade the dataset's utility, affecting the LLM's performance.

    - **Example:** Masking all numerical fields in transaction data may obscure key fraud indicators, reducing model effectiveness.

2. **Compliance with Regulatory Standards:** Different regions and industries have unique requirements for anonymization. Ensuring alignment with these standards adds complexity to data preparation workflows.

    - **Example:** GDPR emphasizes complete anonymization, whereas other regulations may allow pseudonymization.

3. **Re-Identification Risks:** Anonymized data can sometimes be reverse-engineered to re-identify individuals, especially when combined with external datasets.

    - **Example:** A dataset with anonymized location and transaction details could be cross-referenced with public records to infer user identities.

CHAPTER 5   ENSURING DATA PRIVACY AND SECURITY

4. **Computational Overheads:** Techniques like differential privacy involve computational costs, which can impact the scalability of LLM operations.

   - **Example:** Adding noise to millions of records may require significant processing power, delaying data preparation.

As shown in Figure 5-1, data anonymization for LLM training involves techniques like masking, generalization, pseudonymization, and differential privacy. These steps ensure sensitive financial data is protected while still enabling effective model training

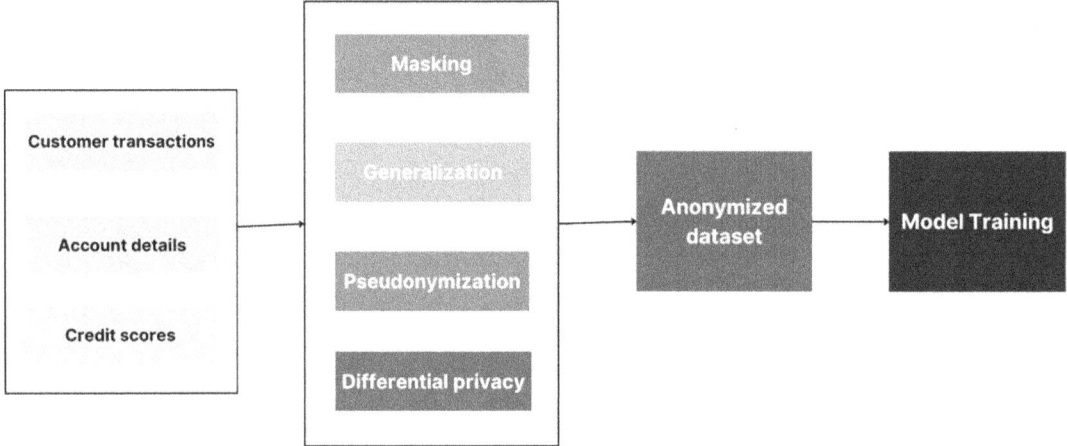

Figure 5-1. Data Anonymization Workflow for LLM Training

Case Study: Fraud Detection

A bank wants to train an LLM to detect fraudulent transactions using customer data. To ensure privacy and compliance, the following steps are taken:

1. **Data Masking:** Account numbers are replaced with pseudonyms, ensuring no direct identifiers remain.

2. **Generalization:** Transaction locations are reduced to city-level details (e.g., "New York" instead of "123 5th Avenue, NY").

3. **Differential Privacy:** Small amounts of noise are added to transaction values, preserving patterns but obscuring exact amounts.

These steps result in a dataset that is both privacy-compliant and suitable for training a high-performing fraud detection LLM. Data anonymization is a critical component of ensuring privacy and compliance in LLM workflows. Techniques like masking, generalization, and differential privacy enable organizations to protect sensitive information while maintaining the utility of datasets for model training and operations. However, balancing privacy with data usability and adhering to regulatory requirements remain ongoing challenges. By adopting best practices and leveraging appropriate anonymization techniques, financial institutions can confidently integrate LLMs into their operations without compromising on privacy or compliance.

Secure Data Storage

Storing sensitive financial data securely is a cornerstone of ensuring data privacy and compliance in LLM workflows. Secure storage methods protect data from unauthorized access, breaches, and misuse, safeguarding both the organization and its customers. This section explores the key principles of secure data storage, strategies tailored to LLM workflows, and best practices for cloud-based storage, with practical examples for financial applications.

In addition to traditional data protection, special attention must be given to **model checkpoints** and **fine-tuning datasets**, especially in **transfer learning** scenarios. These artifacts may contain embedded traces of sensitive information from original training data. As such, organizations should implement **encryption-at-rest**, **access controls**, and **audit trails** for model storage locations—whether using services like **AWS S3 with KMS**, **Azure Blob Storage with RBAC**, or **GCP Cloud Storage with CMEK**—to ensure end-to-end protection of both data and models.

The key principles of secure data storage are as follows:

1. **Encryption Methods:** Encryption transforms readable data into an unreadable format, ensuring that only authorized parties can access it using decryption keys.

   - **Common Methods:**
     - **AES (Advanced Encryption Standard):** A symmetric encryption method widely used for encrypting large volumes of data efficiently.

CHAPTER 5 ENSURING DATA PRIVACY AND SECURITY

- **RSA (Rivest-Shamir-Adleman):** An asymmetric encryption technique often used for secure key exchanges.
- **Example:** Encrypting financial transaction logs with AES before storing them in a da tabase ensures they remain secure even if the database is breached.

2. **Secure Access Controls:** There are two types of data access controls:
    - **Role-Based Access Control (RBAC):** Restricts data access based on user roles and responsibilities, ensuring that employees only access data relevant to their tasks.
    - **Multi-Factor Authentication (MFA):** Adds an additional layer of security by requiring multiple forms of verification, such as a password and a biometric scan.
    - **Example:** Only compliance officers with MFA-enabled accounts can access sensitive customer transaction data stored in encrypted files.

As shown in Figure 5-2, secured data storage is built in layers: encryption at the base, access control in the middle, and cloud storage with audit logging at the top. This layered approach strengthens data protection and compliance.

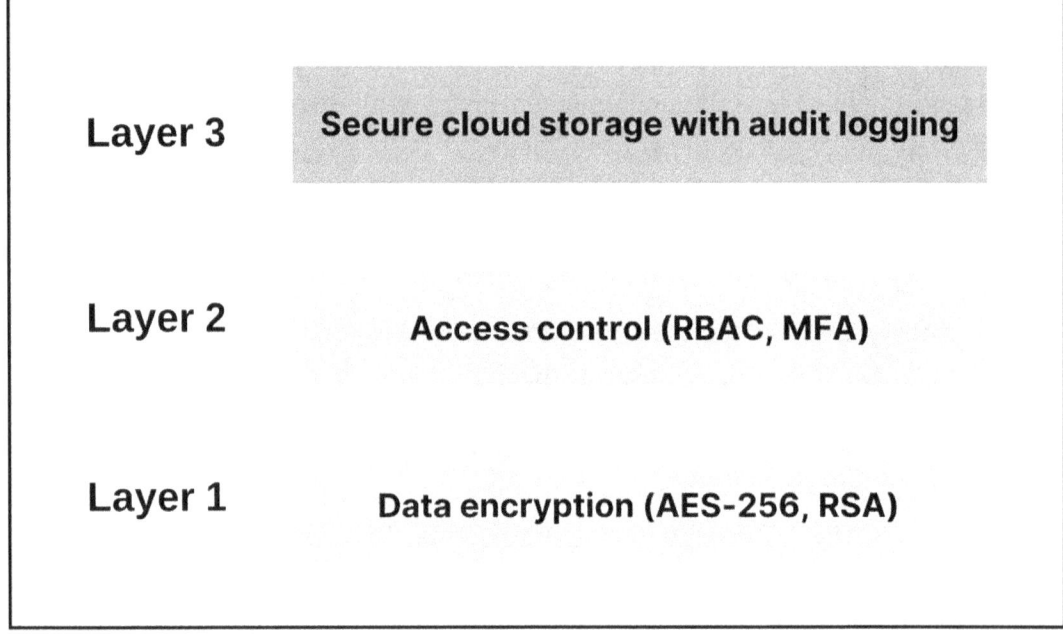

Figure 5-2. Secured Data Storage

As shown in Table 5-2, storage methods like AES, RSA, tokenization, RBAC, and MFA secure financial data through encryption, access control, and authentication. Each method enhances protection against unauthorized access and data breaches.

Table 5-2. *Storage Methods*

Storage Method	Description	Best Use Case	Security Benefits
AES Encryption	Uses symmetric encryption to protect stored data.	Encrypting financial transaction records.	Prevents unauthorized access even if data is stolen.
RSA Encryption	Uses asymmetric keys for secure data exchanges.	Securely transmitting sensitive financial reports.	Ensures that only authorized users can decrypt data.
Tokenization	Replaces sensitive data with a random token.	Protecting credit card numbers in transactions.	Eliminates exposure of actual sensitive data.
Role-Based Access Control (RBAC)	Restricts access based on user roles.	Allowing only auditors to access compliance reports.	Minimizes exposure to unauthorized users.
Multi-Factor Authentication (MFA)	Requires multiple authentication factors.	Logging into a secure LLM training environment.	Reduces risks of credential theft.

The following are the strategies for storing data in LLM workflows:

1. **Partitioning Data by Sensitivity Levels:** Dividing data into categories based on sensitivity and applying appropriate security measures to each category.

    - **Implementation:**

        - **Highly Sensitive Data:** Encrypt and restrict access to data like customer identities and transaction details.

        - **Moderately Sensitive Data:** Use tokenization or pseudonymization for intermediate protection.

        - **Non-Sensitive Data:** Store general metadata with minimal security measures.

- **Example:** In a fraud detection system, customer account details (highly sensitive) are stored separately from anonymized transaction summaries (non-sensitive).

2. **Tokenization:** Replacing sensitive data fields with non-sensitive tokens while maintaining the format of the original data.

    - **Use Case:** Ideal for securely storing PII, such as names or credit card numbers, without exposing the original values.

    - **Example:** Storing tokenized versions of credit card numbers (4111-XXXX-XXXX-1234) in training datasets for an LLM used to analyze purchasing patterns.

3. **Audit and Logging Systems:** Maintaining detailed logs of data access and modifications to track potential breaches or unauthorized activities. Regular audits ensure compliance with internal policies and regulatory requirements.

    - **Example:** Logging all access attempts to transaction data used for LLM training, including user ID, timestamp, and access duration.

Cloud platforms are increasingly popular for storing and processing data used in LLM workflows due to their scalability and cost-efficiency. However, they come with unique challenges that require careful management.

Benefits of Cloud-Based Storage

As financial institutions scale their LLM operations, cloud-based storage provides a flexible, efficient, and cost-effective solution for handling large volumes of financial data while ensuring accessibility and operational continuity.

> **Scalability:** Easily accommodates growing data volumes, essential for LLM operations that require extensive training datasets.
>
> **Accessibility:** Enables secure access from multiple locations, facilitating remote collaborations.
>
> **Cost-Effectiveness:** Reduces infrastructure costs by eliminating the need for on-premise hardware.

Leveraging cloud storage allows organizations to efficiently manage data-intensive LLM workflows while reducing infrastructure overhead and enabling seamless global operations.

Risks of Cloud-Based Storage

While cloud storage offers significant advantages, it also introduces security and dependency risks that must be carefully managed to protect sensitive financial data and ensure regulatory compliance.

> **Data Breaches:** Potential exposure of sensitive data due to misconfigurations or vulnerabilities.
>
> **Vendor Dependency:** Relying on third-party platforms increases risks if the provider fails to maintain adequate security standards.

Understanding these risks allows financial organizations to implement stronger security measures and ensure cloud providers meet the highest compliance and privacy standards.

Best Practices

To mitigate risks and enhance data security, financial institutions must adopt stringent best practices that protect stored data, ensure controlled access, and prevent security breaches.

> **Encryption at Rest and Transit:** Encrypt data stored on cloud servers and data transmitted between systems. Example: Using SSL/TLS protocols for secure data transfer.
>
> **Access Management:** Implement RBAC and enforce MFA for all users accessing the cloud storage.
>
> **Regular Security Audits:** Conduct periodic assessments of cloud configurations and compliance with organizational policies.
>
> **Data Backup and Recovery:** Regularly back up data to prevent loss due to accidental deletions or ransomware attacks.

By adopting these best practices, financial institutions can harness the benefits of cloud-based storage while maintaining the highest standards of data security and regulatory compliance.

The Table 5-3 outlines key practices such as data minimization, audits, access control, and consent management. Each practice is linked to concrete financial examples, ensuring compliance and ethical data handling.

Table 5-3. SBest Practices for Responsible Data Use in LLMs

Best Practice	Description	Example in Finance
Data Minimization	Collect only necessary data for LLM training.	Removing PII from datasets before processing.
Regular Audits	Conduct security and compliance reviews periodically.	Quarterly audits of LLM inference logs.
Access Control Policies	Define strict access rules for different user roles.	Only compliance officers can modify regulatory models.
Consent Management	Allow users to opt out of data collection.	Implementing a GDPR-compliant data deletion process.

CASE STUDY: SECURELY STORING CUSTOMER TRANSACTION RECORDS

A bank trains an LLM to detect fraudulent activities using transaction data from millions of customers. To ensure data privacy and security, the following measures are implemented:

1. **Data Partitioning:** Highly sensitive data, such as customer names and account numbers, is encrypted and stored in a secure vault with limited access. Transaction metadata, such as timestamps and amounts, is stored separately in a pseudonymized format.

2. **Encryption:** All stored data is encrypted using AES-256, ensuring that even if storage systems are breached, the data remains inaccessible.

3. **Cloud Integration:** Customer transaction records are stored on a cloud platform with encryption at rest and transit enabled. RBAC ensures that only authorized data scientists working on LLM training can access anonymized datasets.

4. **Audit Logging:** Every access attempt, data modification, and retrieval request is logged and reviewed periodically.

Outcome:

The bank's data storage strategy protects sensitive information, ensures compliance with regulations like GDPR, and enables efficient LLM training without compromising data security.

Secure data storage is an indispensable component of privacy-first LLM operations. By leveraging advanced encryption techniques, implementing robust access controls, and adopting best practices for cloud-based storage, organizations can safeguard sensitive data against breaches and unauthorized access. These strategies not only ensure regulatory compliance but also build trust with customers, enabling the secure and ethical use of LLMs in financial systems.

Compliance with Financial Regulations

In the financial sector, compliance with data privacy and security regulations is not just a legal obligation but also a critical factor in maintaining trust and credibility. With the advent of large language models (LLMs), which process vast amounts of sensitive data, ensuring compliance has become increasingly complex. This section provides an overview of key financial data regulations, the consequences of non-compliance, strategies for embedding compliance in LLM operations, and practical examples to illustrate best practices.

Understanding Financial Data Regulations

Financial institutions operate under stringent regulatory frameworks to protect sensitive customer data and ensure ethical data handling practices. Some of the most critical regulations affecting LLM operations include the following:

Regulatory Compliance in Large Language Model (LLM) Operations
As the adoption of large language models (LLMs) accelerates across industries—especially in finance, healthcare, and customer service—organizations are facing increasing pressure to comply with complex and evolving regulations around data privacy, security, and accountability. In this chapter, we will explore the regulatory frameworks that shape how LLMs can be designed, trained, and deployed responsibly. We will also present best practices for ensuring compliance while maintaining the power and flexibility of LLMs in business-critical environments.

Key Regulations Shaping LLM Development and Deployment

General Data Protection Regulation (GDPR)

The GDPR is a comprehensive data protection regulation enforced across the European Union. It governs how organizations collect, store, process, and share personal data. The regulation emphasizes user control and transparency, with the following core provisions:

- **Data Minimization**: Organizations must collect only the data necessary for a specific, well-defined purpose. This limits the scope of data collection and reduces risk.

- **Right to Be Forgotten**: Individuals can request the deletion of their personal data. This has profound implications for LLMs, which may need to remove specific information from training datasets or model memory.

- **Data Portability**: Users must have access to their data in a structured, commonly used, and machine-readable format.

Impact on LLMs: To comply with GDPR, training data must be either anonymized or pseudonymized. Identifiable information must not be retained in model outputs or internal representations. Developers must implement differential privacy or other anonymization techniques to minimize re-identification risks. Retrieval-augmented generation and fine-tuned models must be designed to respect user consent and deletion requests.

In the context of LLM deployment, it's also important to distinguish between **data controllers,** entities that determine the purpose and means of processing personal data, and **data processors,** entities that process data on behalf of the controller. When LLMs are fine-tuned by third-party vendors, clear contractual and technical boundaries must be established to uphold GDPR responsibilities, especially in cross-border or multi-party model development scenarios.

California Consumer Privacy Act (CCPA)

The CCPA is a state-level regulation in the United States that gives California residents greater control over their personal data. Although not as comprehensive as the GDPR, it introduces significant requirements for companies that collect or sell personal data:

- **Data Access Rights**: Consumers can request to know what data is being collected about them.

- **Opt-Out Rights**: Users can prevent their data from being sold or used in specific applications.

- **Deletion Requests**: Similar to the GDPR's right to be forgotten, consumers can request data deletion.

Impact on LLMs: Organizations must ensure that personal data included in LLM training or inference respects opt-out choices and deletion requests. This may involve building dynamic opt-out filtering systems into the model pipeline or retraining models on filtered datasets.

Payment Card Industry Data Security Standard (PCI DSS)

PCI DSS is a global standard designed to protect cardholder data. It applies to all entities that store, process, or transmit credit card information and includes the following requirements:

- **Encryption**: Cardholder data must be encrypted both at rest and in transit.

- **Access Controls**: Only authorized personnel should have access to sensitive payment data.

- **Monitoring and Logging**: Systems must track and monitor all access to network resources and cardholder data.

Impact on LLMs: Models trained or used for payment fraud detection must handle payment data in accordance with PCI DSS standards. This includes using secure data ingestion pipelines, encrypting input and output, and limiting model access to verified personnel. Integration with key management systems and access control lists is essential.

Tip Implement robust access controls, such as role-based access control (RBAC) and multi-factor authentication (MFA), to limit access to sensitive data and LLM systems. Also, establish comprehensive logging and monitoring to detect and respond to unauthorized access or suspicious activity.

The High Cost of Non-Compliance

Failure to comply with data regulations can result in substantial penalties and long-term reputational damage. Key consequences include

- **Financial Penalties**:
  - GDPR: Fines up to €20 million or 4% of annual global revenue.
  - CCPA: Up to $2,500 per unintentional violation and $7,500 per intentional violation.
- **Reputational Damage**:
  - Data breaches or regulatory fines can lead to loss of customer trust.
  - Public scrutiny and media attention can affect stock prices and brand loyalty.
- **Operational Disruptions**:
  - Regulatory investigations can result in audits, lawsuits, and temporary suspension of operations.
  - Engineering teams may need to pause innovation to focus on compliance retrofits.

A single breach or violation can trigger multiple forms of damage, ranging from consumer lawsuits to internal morale issues. Companies must treat compliance as a proactive, ongoing discipline rather than a one-time checklist.

Caution Even when using anonymization techniques like masking or generalization, be aware that determined actors may still be able to re-identify individuals by combining anonymized data with other available information.

Embedding Compliance in LLM Workflows

To successfully integrate LLMs into regulated environments, organizations must incorporate compliance into the design and operational lifecycle of their models. Here are three foundational pillars of a compliant LLM system:

1. **Data Minimization Techniques**

   Definition: Collect and process only the data necessary to achieve defined objectives.

   Implementation Steps:

   - Identify and classify data based on sensitivity and use case.
   - Apply anonymization or pseudonymization to reduce exposure
   - Limit collection of personal identifiers and linkages.

   Example: For an LLM trained on financial transaction patterns, customer names and account numbers should be removed, while transactional metadata (e.g., time, amount, location) is retained.

2. **Conducting Regular Audits**

   Purpose: Maintain continuous compliance and adapt to evolving regulations.

   Implementation Steps:

   - Conduct quarterly audits of data pipelines and access logs.
   - Validate that anonymization and data retention policies are followed.
   - Use automated monitoring tools to detect unauthorized access or data leaks.

   Example: An audit of a credit scoring LLM may include a review of training datasets, evaluation of fairness metrics, and validation of GDPR-compliant data retention practices.

3. **Maintaining Clear Documentation**

   Definition: Keep detailed records of how data is collected, processed, stored, and shared.

   Benefits:

   - Facilitates internal reviews and external audits.
   - Demonstrates due diligence and accountability to regulators.
   - Serves as a blueprint for future model iterations.

Example: Documentation for an LLM used in customer support should include a data flow diagram, privacy impact assessment, data anonymization methods, and a changelog of model updates.

As LLMs become core components of decision-making in high-stakes environments, regulatory compliance cannot be an afterthought. By embracing data minimization, regular auditing, and transparent documentation, organizations can mitigate risk while fostering innovation. Aligning with GDPR, CCPA, and PCI DSS from the ground up enables companies to deploy LLMs with confidence, earning user trust and regulatory approval in equal measure.

Tip   Integrate data privacy and security considerations into every stage of the LLM lifecycle, from data collection and preparation to model training, deployment, and monitoring. This proactive approach, known as "Privacy by Design," helps prevent issues down the line and is more effective than trying to retrofit security measures.

Example Practices

Integrating privacy regulations into LLM operations requires proactive measures to ensure compliance with laws like GDPR's "Right to Be Forgotten" clause. Financial institutions must implement structured data management strategies to respect user privacy while maintaining AI model integrity.

1. Aligning LLM Operations with GDPR's "Right to Be Forgotten" Clause

    - **Challenge**: Ensuring that personal data used for training LLMs is deleted upon customer request.

    - **Solution**: Maintain a mapping between raw data and anonymized datasets to enable targeted deletions. Implement training workflows that allow incremental updates and retraining models without retaining non-compliant data.

    - **Outcome**: The organization avoids fines and maintains customer trust by respecting deletion requests.

2. Ensuring Third-Party Vendor Compliance

    - **Challenge**: Many organizations rely on third-party vendors for data storage, processing, or model development.

    - **Solution**: Vet vendors thoroughly to ensure their compliance with GDPR, CCPA, and PCI DSS. Include compliance clauses in contracts and conduct periodic audits.

    - **Example**: A bank collaborates with a cloud storage provider that encrypts data and complies with regional privacy laws, ensuring secure storage of training datasets.

Compliance with financial data regulations is a critical aspect of LLM operations. By understanding the requirements of key regulations such as GDPR, CCPA, and PCI DSS, organizations can implement strategies like data minimization, regular audits, and clear documentation to ensure adherence. Real-world practices, such as enabling customer data deletion and vetting third-party vendors, highlight the practical steps required to embed compliance in LLM workflows. Proactively addressing regulatory challenges not only safeguards against penalties but also reinforces customer trust and operational integrity in the dynamic financial landscape.

Caution LLMs are often accessed via APIs, which can become a significant vulnerability if not properly secured. Weak authentication, lack of rate limiting, and insufficient input validation can expose sensitive data to unauthorized parties.

Challenges and Best Practices

Implementing secure and compliant LLM workflows in the financial sector comes with its own set of challenges. Organizations must navigate the complexities of balancing data utility with privacy, addressing emerging threats, and staying ahead of regulatory changes. Financial institutions operate under stringent regulatory frameworks to protect sensitive customer data and ensure ethical data handling practices. This section explores key challenges in implementing secure and compliant LLM workflows and outlines best practices to mitigate risks, ensuring robust and secure operations.

Key Challenges

In the financial sector, compliance with data privacy and security regulations is a critical factor in maintaining trust and credibility. With the advent of large language models (LLMs), which process vast amounts of sensitive data, ensuring compliance has become increasingly complex. Financial institutions must contend with several key regulations, such as the General Data Protection Regulation (GDPR), California Consumer Privacy Act (CCPA), and Payment Card Industry Data Security Standard (PCI DSS). These regulations enforce strict rules on how personal data is collected, processed, and stored, and non-compliance can lead to severe consequences.

For example, under the GDPR, organizations are required to anonymize or pseudonymize training datasets to prevent unauthorized access to personal data. Similarly, the CCPA emphasizes transparency and consumer rights, requiring that customers have control over how their data is used, including the right to opt out. For financial institutions, complying with regulations like PCI DSS is paramount, as it requires encrypting sensitive cardholder data during storage and transmission. Failure to comply with these regulations can result in substantial fines, reputational damage, and operational disruptions.

Adversarial Robustness and Model Red Teaming

One growing challenge is defending LLMs against adversarial attacks, where subtle inputs are crafted to mislead the model. To address this, institutions should conduct model red teaming exercises to uncover vulnerabilities in real-world scenarios. Open-source tools like IBM's Adversarial Robustness Toolbox and Microsoft's Counterfit offer practical means to simulate such threats and evaluate the model's resilience. These tools help teams test LLM behavior under adversarial prompts, manipulation attempts, or privacy extraction risks.

Auditability and Compliance Logging

Another best practice is ensuring end-to-end auditability of LLM workflows. Audit logs should capture model access, training events, inference records, and administrative actions in a secure and tamper-proof format. Aligning audit trail mechanisms with SOC 2 or ISO/IEC 27001 requirements ensures that the LLM system adheres to industry standards for security, integrity, and traceability—crucial for regulatory audits and internal governance.

Consequences of Non-Compliance

Failure to comply with financial data regulations can have far-reaching consequences. Regulatory violations can lead to heavy fines, with penalties such as €20 million or 4% of global annual revenue for GDPR violations, or up to $7,500 per violation under CCPA. Financial penalties can cripple an organization's bottom line, but the damage is not limited to fines alone. Reputational harm is another significant risk, as publicized non-compliance incidents can erode customer trust, leading to market devaluation and customer attrition. Rebuilding a tarnished reputation is a complex and costly process. Furthermore, non-compliance can cause operational disruptions as investigations, audits, and legal actions may freeze business operations or disrupt workflows, making it essential to maintain ongoing compliance.

Best Practices for Ensuring Compliance

To ensure compliance and mitigate risks, financial institutions must adopt proactive measures that embed data protection strategies within LLM workflows. Some of the best practices to follow include

1. **Data Minimization Techniques**: Limiting data collection and processing to only what is necessary ensures compliance with privacy laws while reducing exposure to security risks. Financial institutions should identify and classify data based on sensitivity and necessity, removing unnecessary personal identifiers from training datasets. For example, when training an LLM for financial transaction pattern recognition, anonymizing customer identifiers while retaining transaction metadata can ensure compliance with data privacy regulations.

2. **Conducting Regular Audits**: Regular audits of data handling practices are essential to ensure compliance with evolving regulations and identify potential vulnerabilities. Audits should review access logs to ensure only authorized personnel access sensitive data, validate anonymization techniques, and rectify vulnerabilities in data pipelines. For example, a quarterly audit of an LLM-powered credit scoring system could confirm that the training data complies with GDPR's anonymization standards.

3. **Maintaining Clear Documentation**: Comprehensive documentation of data collection, storage, processing, and sharing practices enhances accountability and transparency in financial LLM operations. It facilitates regulatory audits and internal reviews and demonstrates the organization's commitment to regulatory compliance. Documenting how personal data is anonymized and stored during LLM training helps create a clear audit trail and ensures regulatory transparency.

Tip Data privacy and security regulations are constantly evolving, with new laws and interpretations emerging. Establish a process for staying informed about these changes and adapting your LLM operations accordingly.

As financial organizations scale their LLM deployments, they must continue to balance privacy with data utility, mitigate adversarial risks, and ensure compliance with evolving regulations. By following these best practices, institutions can safeguard sensitive customer data, enhance operational transparency, and minimize risks associated with regulatory violations, maintaining trust and credibility in their LLM operations.

Emerging Technologies

As financial institutions seek more secure and privacy-preserving solutions for deploying LLMs, emerging technologies such as homomorphic encryption and federated learning are reshaping the landscape of AI-driven financial operations. These innovations offer enhanced security, reduced data exposure risks, and improved compliance with evolving privacy regulations. By integrating these technologies, organizations can develop more secure, efficient, and collaborative AI models while maintaining strict control over sensitive financial data.

Homomorphic Encryption

Homomorphic encryption is a sophisticated cryptographic technique that allows computations to be carried out directly on encrypted data without the need to decrypt it. This ensures that the data remains secure even during processing, significantly reducing the risk of exposing sensitive information. In the context of large language models (LLMs), homomorphic encryption enables secure training on encrypted datasets, where raw financial data never needs to be revealed. It also supports encrypted inference, allowing queries to be processed without exposing user input. A practical example in finance is a bank using this technique to detect fraudulent credit card transactions without accessing the actual transaction details. Despite its strong privacy advantages, homomorphic encryption faces challenges such as high computational overhead and complex implementation, making it difficult to deploy in real-time or large-scale systems without advanced infrastructure and expertise.

A practical example in finance is a bank using this technique to detect fraudulent credit card transactions without accessing the actual transaction details. Despite its strong privacy advantages, **homomorphic encryption** faces challenges such as high computational overhead and complex implementation, making it difficult to deploy in real-time or large-scale systems without advanced infrastructure and expertise. Currently, its use is largely limited to **niche scenarios**, such as **encrypted inference on sensitive data**, rather than full-scale LLM training, due to performance bottlenecks.

Federated Learning

Federated learning is a distributed machine learning approach that enables models to be trained across multiple devices or institutions without transferring raw data to a central location. Instead of pooling data in a single server, the model learns locally at each node, preserving data privacy and minimizing the risks of centralized data storage. This makes it particularly well-suited for financial services, where sensitive customer data must be handled with strict confidentiality. For large language models, federated learning allows institutions to collaborate on model training—such as for fraud detection or anti-money laundering—without sharing proprietary or personal data. It also enables on-device learning, where LLMs are updated using data from customer devices like mobile banking apps, without uploading that data to the cloud. A real-world example includes a consortium of banks training a shared LLM-based AML model using anonymized, locally stored data. While federated learning promotes secure and privacy-preserving AI development, it does require strong infrastructure for securely aggregating model updates and may not always match the performance of centralized approaches.

The integration of **homomorphic encryption and federated learning** marks a **transformative shift** in how financial institutions handle LLM security and compliance. These technologies **offer promising solutions to data privacy challenges** while enabling **collaborative AI development** without compromising sensitive information.

As adoption increases and infrastructure matures, these innovations will play a crucial role in **shaping the future of secure and privacy-first LLM deployments** in the financial sector.

Evolving Regulations

As privacy concerns continue to grow, governments and regulatory bodies worldwide are adopting stricter data protection laws to prevent misuse and strengthen consumer rights. These anticipated changes are expected to impose enhanced transparency requirements, stricter controls over cross-border data transfers, and increased accountability for AI-driven decision-making. For organizations leveraging LLMs, this evolving regulatory landscape necessitates proactive measures to ensure compliance across multiple jurisdictions.

In the European Union, amendments to GDPR may introduce stricter explainability and accountability requirements for AI systems, ensuring that financial institutions can justify model decisions in areas like credit scoring and fraud detection. In the United States, the California Consumer Privacy Act (CCPA) is expanding, with additional states such as Virginia and Colorado introducing their own privacy laws, leading to a more fragmented regulatory environment. Meanwhile, Asia-Pacific nations, including India and Singapore, are updating their privacy laws to align with international standards, which could impact multinational financial institutions deploying LLMs across different regions. As these regulations evolve, financial organizations must ensure that LLM workflows comply with region-specific laws, adding complexity to cross-border AI operations and increasing the need for comprehensive documentation and audit trails.

Regulatory enforcement is also expected to become more stringent, with financial authorities focusing on holding organizations accountable for non-compliance. This shift places a stronger emphasis on proactive compliance measures, such as regular audits, maintaining transparent documentation, and implementing robust incident response plans to address potential breaches. LLM operations will need to prioritize explainability, ensuring that AI-driven decisions, particularly in areas like loan approvals and fraud detection, are interpretable and auditable.

LLM operations will need to prioritize explainability, ensuring that AI-driven decisions, particularly in areas like loan approvals and fraud detection, are interpretable and auditable. Under anticipated **EU AI Act** provisions and related financial regulations, explainability for credit scoring models may become **legally mandated**. This will require LLM systems to **log inference traces** or support **counterfactual explanations** that clarify how a specific outcome was reached. These explainability features help satisfy compliance, facilitate audits, and build trust with regulators and end-users.

Additionally, consent management frameworks will become a critical aspect of AI governance, requiring organizations to implement mechanisms that allow users to provide, withdraw, and track consent for data usage in LLM workflows.

To mitigate risks and stay compliant, financial institutions are beginning to integrate real-time compliance monitoring into their LLM workflows. These monitoring systems track data flows, access permissions, and model outputs to detect and flag potential privacy violations before they escalate. For example, a financial institution utilizing LLMs for transaction monitoring may deploy real-time security alerts to identify unauthorized access attempts to sensitive data. By embedding continuous compliance checks into AI-driven operations, organizations can reduce regulatory risks, ensure ethical AI usage, and build customer trust in an increasingly regulated financial ecosystem.

Implications for Organizations

As data privacy regulations evolve and cybersecurity threats become more sophisticated, financial institutions must adopt proactive strategies to ensure LLM security, compliance, and operational resilience. Organizations that integrate future-proof technologies, strengthen regulatory collaboration, and build adaptable compliance frameworks will be better positioned to navigate the complexities of AI-driven financial ecosystems.

1. **Adopting Future-Proof Technologies:**

    - Implementing homomorphic encryption and federated learning allows organizations to meet stringent privacy requirements while maintaining operational effectiveness.

2. **Enhanced Collaboration with Regulators:**

    - Establishing transparent communication with regulatory bodies ensures alignment with evolving standards and fosters trust.

3. **Building Resilient Compliance Frameworks:**

    - Regularly updating compliance protocols to address new regulations and emerging threats is essential for sustainable operations.

By proactively adopting secure technologies, fostering regulatory collaboration, and maintaining adaptive compliance frameworks, organizations can future-proof their LLM operations while safeguarding financial data and ensuring trust in AI-driven decision-making.

Conclusion

Ensuring data privacy and security in large language model (LLM) operations is not just a regulatory requirement but a fundamental pillar of responsible AI deployment in finance. As financial institutions increasingly leverage LLMs for fraud detection, risk assessment, and customer service, protecting sensitive data from breaches, adversarial attacks, and compliance violations is critical. This chapter has provided a comprehensive roadmap for securing LLM workflows, covering data anonymization techniques, encryption strategies, regulatory compliance frameworks, and best practices for mitigating emerging threats.

By implementing privacy-preserving techniques such as masking, generalization, and differential privacy, organizations can balance security with data utility, ensuring compliance with stringent regulations like GDPR, CCPA, and PCI DSS. Secure storage mechanisms, including AES and RSA encryption, role-based access controls (RBAC), and multi-factor authentication (MFA), further safeguard financial data against unauthorized access. Additionally, integrating real-time monitoring, automated audits, and regulatory reporting helps organizations maintain compliance while adapting to evolving legal landscapes.

Emerging trends, such as homomorphic encryption and federated learning, offer new pathways for enhancing privacy-first LLM workflows. As global regulations tighten and cyber threats grow more sophisticated, financial institutions must proactively adopt future-proof security strategies, collaborate with regulators, and build resilient compliance frameworks to ensure long-term AI sustainability.

Points to Remember

- **Data Privacy Is Non-negotiable:** Protecting customer and financial data is essential for regulatory compliance, trust, and ethical AI operations.

- **Regulatory Adherence Is Key:** Regulations like **GDPR, CCPA, and PCI DSS** require strict compliance, including data minimization, access controls, and auditing.

- **Data Anonymization Enhances Security:** Techniques such as **masking, suppression, and differential privacy** ensure that personally identifiable information (PII) is protected while maintaining model effectiveness.

- **Encryption Is Essential for Secure Storage:** Using **AES-256, RSA encryption, and end-to-end data encryption** prevents unauthorized access to sensitive datasets.

- **Access Control and Authentication Strengthen Security:** Implementing **RBAC, MFA, and continuous monitoring** limits unauthorized access to critical financial data.

- **Monitoring and Auditing Ensure Compliance:** Organizations should establish **real-time tracking, anomaly detection, and regular audits** to prevent security breaches.

- **Emerging Threats Must Be Addressed:** Adversarial attacks, **data poisoning, and model inversion risks** necessitate proactive security updates and model hardening.

- **Cloud Security Requires Robust Safeguards:** Encrypting data in transit and at rest, **vetting third-party vendors**, and enforcing strict access controls minimize cloud storage risks.

- **Collaboration with Legal and Compliance Teams Is Crucial:** A cross-functional approach ensures LLM deployments align with legal frameworks and industry best practices.

- **Future-Proof Technologies Enhance Security:** Leveraging **homomorphic encryption, federated learning, and automated compliance monitoring** prepares organizations for stricter regulations and evolving cybersecurity threats.

CHAPTER 6

Integrating LLMs into Financial Systems

Integrating large language models (LLMs) into existing financial systems requires careful planning to ensure compatibility, performance, and compliance with operational standards. This chapter provides strategies and best practices for seamlessly incorporating LLMs into financial infrastructures, enabling smooth data flow and interoperability across platforms. The chapter begins with an exploration of API development and management, guiding you through the process of creating and managing RESTful APIs that facilitate efficient communication between LLMs and other system components.

Real-time data processing is another critical aspect of integration, especially for finance applications that rely on fast, accurate insights. This chapter covers the essentials of handling streaming data and integrating LLMs with real-time systems, optimizing performance for time-sensitive tasks. Finally, case studies of successful LLM integrations in finance offer practical insights and examples, illustrating how other organizations have effectively implemented LLMs to enhance their financial systems. By mastering these integration techniques, you will be able to enhance the capabilities of financial systems while maintaining stability and operational efficiency.

Structure

This chapter covers the following topics:

- **API Development and Management:**

    Guidelines for designing and managing APIs to facilitate seamless interaction between LLMs and financial systems. Focus areas include building scalable and secure RESTful APIs, versioning for continuous updates, and ensuring backward compatibility to maintain system stability.

- **Real-Time Data Processing:**

  Techniques for integrating LLMs with real-time systems in finance, including handling streaming data with tools like Apache Kafka and ensuring low-latency responses. This section covers architectural considerations, scalability strategies, and managing throughput for high-demand applications.

- **Case Studies:**

  Examples of successful LLM integrations in financial systems, demonstrating how challenges were addressed and solutions implemented. Case studies include LLM-powered customer service chatbots, real-time market sentiment analysis, and automated compliance reporting systems.

Objectives

At the end of this chapter, you will be equipped with the knowledge and tools to integrate LLMs effectively into existing financial systems. You will understand how to develop and manage APIs that ensure smooth communication between LLMs and other system components, including techniques for maintaining version control and compatibility. You will also gain skills in real-time data processing, allowing you to handle streaming data and incorporate LLMs into high-performance systems that meet the demands of financial applications. Through real-world case studies, you will gain insights into the challenges and solutions involved in LLM integration, learning best practices that can be applied in your own implementations. With these integration strategies, you will be able to enhance the capabilities of financial systems, supporting operational efficiency and driving innovation.

API Development and Management

APIs (application programming interfaces) play a pivotal role in integrating large language models (LLMs) with financial systems. They serve as the communication bridge that allows financial applications to interact seamlessly with LLMs for tasks such

CHAPTER 6   INTEGRATING LLMS INTO FINANCIAL SYSTEMS

as fraud detection, customer support, and regulatory compliance. Effective API design and management ensure scalability, security, and maintainability, which are crucial in the complex and regulated financial sector.

This section provides detailed guidelines for designing and managing APIs tailored for LLM integration, focusing on building scalable and secure RESTful APIs, implementing versioning for updates, and ensuring backward compatibility to maintain system stability.

Key Guidelines for API Design

APIs are the backbone of LLM integration in financial systems, enabling seamless communication between services. This section outlines essential principles for designing scalable, secure, and maintainable RESTful APIs tailored for high-performance financial applications.

1. **Building Scalable RESTful APIs:** RESTful APIs (representational state transfer) use standard HTTP methods (GET, POST, PUT, DELETE) to facilitate communication between clients and servers. They are widely used for their simplicity, scalability, and compatibility with modern applications.

   Design Principles:

   - **Resource-Oriented Architecture:** Structure the API around logical resources (e.g., /fraud-check, /credit-score).

   - **Statelessness:** Each API request contains all the information needed to process it, improving scalability.

   - **Pagination and Rate Limiting:** Handle large data sets and prevent server overload by limiting the number of requests per second.

   Example in Finance: A credit scoring system API exposes endpoints like /credit-score to fetch credit evaluations for customers based on their data.

2. **Ensuring Security**

    - **Authentication and Authorization:**
        - Use token-based authentication mechanisms like OAuth 2.0 or API keys to secure endpoints.
        - Implement role-based access controls to restrict API usage based on user roles.
        - **Incorporate token expiration and refresh strategies** to maintain secure statelessness and reduce risk from stolen or reused tokens.
    - **Encryption:**
        - Use HTTPS for encrypted communication to protect sensitive financial data.
    - **Input Validation:**
        - Validate all inputs to prevent injection attacks or malformed requests.

    Example in Finance: An API for a compliance reporting system uses OAuth 2.0 to authenticate users and encrypts data transfer using TLS.

3. **Optimizing for Low Latency:** In financial systems, low-latency APIs are critical for real-time applications like trading and fraud detection.

    Best Practices:
    - Minimize API response payloads by sending only essential data.
    - Use caching mechanisms to store frequently accessed data temporarily.
    - Optimize database queries to reduce processing time.

    Example in Finance: A trading system API caches commonly used market data to provide instant responses for time-sensitive decisions.

CHAPTER 6 INTEGRATING LLMS INTO FINANCIAL SYSTEMS

As shown in Table 6-1, security measures in API communication include encryption, authentication, and input validation. These protect data in transit, validate users, and prevent malicious attacks.

Table 6-1. *Security Measures in API Communication*

Measure	Implementation Tool	Purpose
Encryption	HTTPS/TLS	Data-in-transit protection
Authentication	OAuth 2.0/JWT	User Validation
Input Validation	Schema validators (e.g., JSON schema)	Prevent Attacks

Caution   Do not over-rely on caching, as cached data can become outdated quickly in volatile financial systems. Balance speed with accuracy.

Versioning for Continuous Updates

APIs evolve over time to support new features or improve functionality. Versioning ensures that updates do not disrupt existing integrations, maintaining system stability.

Example in Finance: An API for a loan approval system introduces a new endpoint for calculating dynamic interest rates without breaking existing functionality.

Best Practices for API Versioning

1. **URL Versioning:**

   - Include the version number in the API URL (e.g., /v1/fraud-check, /v2/fraud-check).

   - Makes it easy to identify and manage versions.

   - **Follow semantic versioning** practices (e.g., v1.0, v1.1, v2.0) to clearly indicate whether updates are non-breaking (minor) or breaking (major) changes. This helps consumers of the API understand the impact and manage transitions gracefully.

2. **Header Versioning:**

   - Pass the version number in the API header for cleaner URLs.

171

3. **Deprecation Policy:**

    - Inform users well in advance before deprecating older API versions.

    - Provide a transition period to allow smooth migration to newer versions.

Example Workflow:

- An API for a financial chatbot initially provides basic customer query handling (/v1/chatbot).

- In version 2 (/v2/chatbot), advanced features like sentiment analysis are added without disrupting existing clients.

Tip Use API gateways for control; implement API gateways to manage authentication, traffic throttling, and request routing effectively.

Ensuring Backward Compatibility

Backward compatibility ensures that updates to an API do not break or disrupt existing integrations. Essential for maintaining trust and stability in financial systems where reliability is critical.

Strategies for Maintaining Backward Compatibility

1. **Avoid Breaking Changes:**

    - Do not remove or modify existing endpoints or request/response formats.

2. **Additive Changes Only:**

    - Introduce new features as optional parameters or endpoints to avoid affecting existing users.

3. **Graceful Deprecation:**

    - Mark outdated endpoints as deprecated but functional, allowing time for clients to transition.

CHAPTER 6 INTEGRATING LLMS INTO FINANCIAL SYSTEMS

Example in Finance: A compliance API adds an optional risk-level parameter in its response to enhance functionality while preserving the original structure.

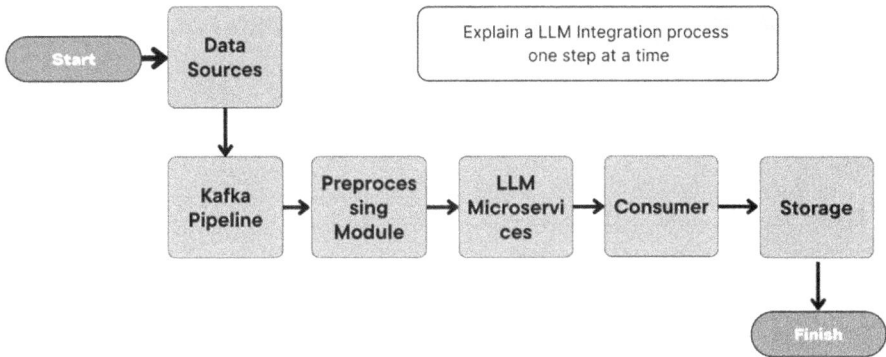

Figure 6-1. *LLM Integration Architecture in Financial Systems*

Figure 6-1 illustrates the **end-to-end integration flow** of large language models (LLMs) into modern financial systems. It demonstrates how real-time data streams, API services, and backend systems work together to support intelligent, scalable AI applications in finance. Below is a breakdown of each component and how they interact:

1. Data Sources

These are the origin points of raw financial and customer data, which can include:

- **Transaction Systems**: Real-time payment and banking transactions.

- **CRM Systems**: Customer profile data, historical interactions, and service logs.

- **Social Media Feeds/News Feeds**: External sources for market sentiment or public events.

These sources continuously generate structured and unstructured data relevant to financial operations.

2. Kafka Pipeline

Apache Kafka serves as the **streaming backbone** for the system:

- **Topics** represent specific data streams (e.g., transactions, market news).

- Data is **partitioned and replicated**, ensuring fault tolerance and scalability.

- Kafka acts as a **decoupler** between producers (data sources) and consumers (LLMs and other processing systems).

Note While Apache Kafka is widely used for streaming pipelines, alternative platforms like **Apache Pulsar** and **Redpanda** also offer high-throughput, low-latency messaging. For example, Pulsar supports native multi-tenancy and geo-replication, while Redpanda is Kafka-compatible with lower latency for certain workloads.

3. Preprocessing Module

Before data reaches the LLMs, it passes through a **preprocessing layer**:

- **Cleansing**: Removes noise, duplicates, and irrelevant records.

- **Enrichment**: Adds contextual metadata (e.g., risk score, location).

- **Formatting**: Transforms data into LLM-compatible input structures (e.g., JSON, plain text, tokenized inputs).

This ensures high-quality, consistent inputs for LLM inference.

4. LLM Microservices

LLMs are deployed as **independent microservices** accessible through RESTful APIs:

- Each LLM serves a specific function (e.g., fraud detection, credit scoring, chatbot).

CHAPTER 6   INTEGRATING LLMS INTO FINANCIAL SYSTEMS

- APIs accept preprocessed data, run inference, and return structured outputs.

- Microservices architecture allows for **easy scaling and modular updates** without affecting the broader system.

Figure 6-2 depicts a **microservices-based architecture** that integrates an LLM (large language model) with a compliance engine. Each functional unit (scoring, preprocessing, reporting, etc.) is implemented as an independent microservice, connected via **RESTful APIs**.

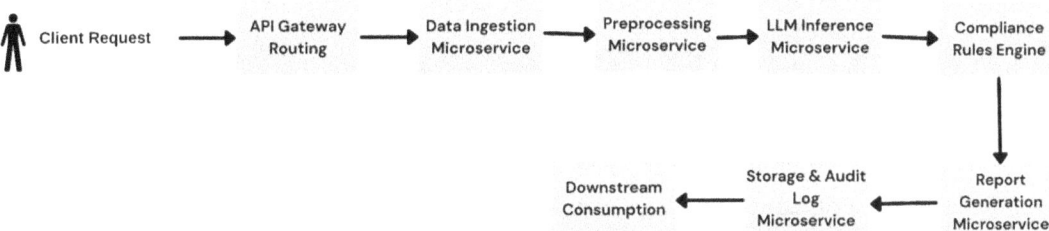

Figure 6-2. *Microservices Architecture with LLM*

Step-by-Step Process Flow

Step 1: Client Request

- A user (e.g., a compliance officer or system agent) makes a request through a **UI or API Gateway**.

- Example: "Generate a risk report for recent transactions."

Step 2: API Gateway Routing

- The request is authenticated and routed by the **API Gateway**.

- The gateway ensures secure, rate-limited access and forwards the request to the correct microservice (e.g., ingestion service).

175

Step 3: Data Ingestion Microservice

- This microservice pulls or receives raw data from
    - Internal databases
    - Transaction feeds
    - Customer relationship management (CRM) systems
- It structures the data and passes it to the **Preprocessing Service**.

Step 4: Preprocessing Microservice

- Tasks include
    - Data cleaning (handling missing values, duplicates)
    - Formatting and normalization
    - Tagging with metadata (e.g., timestamps, customer ID)
- The preprocessed data is forwarded to the **LLM Service**.

Step 5: LLM Inference Microservice

- Receives structured input and performs:
    - Risk scoring
    - Entity extraction
    - Text summarization
- Sends outputs (e.g., fraud scores, compliance summaries) to the **Compliance Engine**.

Step 6: Compliance Rules Engine

- Applies regulatory rules, logic checks, and thresholds.
- Flags any violations or suspicious patterns based on legal requirements.
- Outputs a **compliance verdict** and recommendations.

Step 7: Report Generation Microservice

- Aggregates all results (inference + compliance decision).
- Formats the report in required layouts (PDF, XML, HTML).
- Prepares the report for review, download, or submission to auditors.

Step 8: Storage & Audit Log Microservice

- Stores results in a **centralized repository**.
- Maintains audit trails, model inputs/outputs, and user interaction logs.
- Enables reprocessing or traceability for audits.

Step 9: Downstream Consumption

- Final results are delivered to
    - Dashboards (for visualization)
    - Email/alert systems (for notifications)
    - External systems via APIs

5. Consumers

These are the **front-end applications or systems** that use LLM outputs:

- **Dashboards**: Visualize LLM insights for traders, analysts, or compliance officers.
- **Alerts/Notifications**: Real-time fraud alerts or risk warnings.
- **Customer Service Interfaces**: Chatbots or virtual assistants integrated into banking apps or websites.

Consumers turn LLM outputs into **actionable intelligence**.

6. Storage

Persistent data storage is used for

- **Logging**: Captures all transactions and inferences for audit trails.
- **Historical Database**: Archives input/output pairs for retraining, compliance, and reporting.
- **Compliance Archive**: Stores reports and flagged transactions required for legal and regulatory purposes.

This component supports **traceability**, **model improvement**, and **regulatory alignment**.

Example: API Design for Fraud Detection

A financial institution implements an API to integrate an LLM-based fraud detection system with its transaction monitoring software:

Endpoints:

- /fraud-check: Accepts transaction data and returns a fraud risk score.
- /fraud-log: Logs flagged transactions for auditing purposes.

Security Features:

- Uses OAuth 2.0 for authentication.
- Encrypts data transfer with HTTPS.

Versioning and Compatibility:

- Version 1 supports basic fraud checks.
- Version 2 introduces detailed analysis with additional input parameters (risk level, geo-location).

Outcome:
The API ensures smooth integration, real-time fraud detection, and compliance with data security standards.

API development and management are critical for integrating LLMs into financial systems effectively. By focusing on scalable RESTful designs, robust security measures, versioning for updates, and backward compatibility, organizations can ensure seamless and stable interactions between LLMs and other components. These guidelines enable financial institutions to maximize the utility of LLMs while maintaining the reliability and compliance required in this high-stakes sector.

Tip Deploy LLMs as microservices, as this enhances modularity and makes it easier to independently scale, update, or replace components.

Real-Time Data Processing

Real-time data processing is critical for integrating large language models (LLMs) into financial systems that require instantaneous analysis and decision-making. Applications such as fraud detection, market sentiment analysis, and real-time credit risk assessment demand low-latency responses and the ability to process large volumes of data with high throughput. This section explores the techniques, tools, and strategies necessary to enable seamless integration of LLMs with real-time financial systems, covering streaming data processing, architectural considerations, scalability strategies, and throughput management.

Importance of Real-Time Data Processing in Finance

In the financial sector, decisions are often made in milliseconds. Delays in processing data can result in missed opportunities, financial losses, or compliance risks. Real-time data processing enables:

- **Fraud Detection:** Identifying and flagging suspicious transactions as they occur.

- **Market Analysis:** Providing up-to-the-second insights for algorithmic trading strategies.

- **Customer Interaction:** Powering chatbots and virtual assistants with instant responses.

For example, an LLM-integrated fraud detection system must analyze transaction data streams in real time to block fraudulent activities before they are completed.

CHAPTER 6   INTEGRATING LLMS INTO FINANCIAL SYSTEMS

Techniques for Real-Time Data Processing

Streaming platforms like Apache Kafka and Apache Pulsar enable the continuous ingestion and processing of real-time data. These tools act as intermediaries between the data source and the LLM.

How It Works:
Data producers (e.g., transaction systems, market feeds) send data to Kafka topics. Consumers (e.g., LLM applications) process this data in real time and generate insights.

Example in Finance:
A stock trading platform uses Kafka to stream live market data to an LLM, which provides sentiment analysis for trading decisions.

Low-Latency Inference:

- **Optimizing Model Architecture:** Use smaller or optimized versions of LLMs (e.g., distilled models) to reduce latency.

- **Caching Responses:** Store frequently accessed predictions to serve repeat queries instantly.

Tip   Use Kafka for decoupling. Apache Kafka acts as a reliable event buffer that decouples LLMs from upstream financial systems.

   Example: A customer service chatbot caches responses for common queries, reducing the need for repeated LLM inference.

- **Preprocessing Pipelines**

    Preprocess data in real time to clean, normalize, and enrich it before feeding it into the LLM.

    Example in Finance: A credit scoring model processes raw customer data streams by removing anomalies and formatting input for the LLM.

Architectural Considerations

Event-Driven Architecture: Real-time systems benefit from event-driven architectures that trigger LLM processes only when new data arrives.

Components:

- **Producers:** Systems that generate events (e.g., payment gateways, trading platforms).
- **Message Brokers:** Tools like Kafka that relay events to consumers.
- **Consumers:** LLM-based applications that process and act on the data.

Example: An anti-money laundering system triggers LLM inference whenever a high-risk transaction is detected.

Microservices for Modularity: Deploy LLMs as independent microservices to integrate seamlessly with other system components.

Note Event-driven architecture enables flexibility as trigger-based processing suits LLM integration well, especially in anomaly detection and transaction monitoring.

Benefits:

- **Scalability:** Each microservice can scale independently.
- **Flexibility:** Easily update or replace the LLM without affecting the entire system.

Example: A fraud detection API is a standalone microservice that integrates with payment systems.

Edge Computing for Real-Time Processing: Deploy models on edge devices to process data locally, reducing latency and dependency on centralized systems.

Example: A payment terminal equipped with an edge-deployed LLM detects anomalies before approving a transaction.

CHAPTER 6 INTEGRATING LLMS INTO FINANCIAL SYSTEMS

Scalability Strategies

Horizontal Scaling: Add more instances of LLM inference services to handle increased demand.

Example: During Black Friday sales, a financial institution scales its LLM-powered recommendation engine to accommodate a surge in transaction queries.

Load Balancing: Distribute incoming requests across multiple servers to prevent **bottlenecks.**

Example: A trading platform uses load balancers to ensure that real-time market sentiment analysis is evenly distributed across LLM instances.

Partitioning and Parallelism: Partition data streams (e.g., Kafka topics) and process them in parallel to improve throughput.

Example: A fraud detection system partitions transaction data by geographical region, enabling parallel processing by multiple LLM instances.

Tip Benchmark latency regularly and continuously to measure the end-to-end latency from data ingestion to LLM inference to ensure real-time performance.

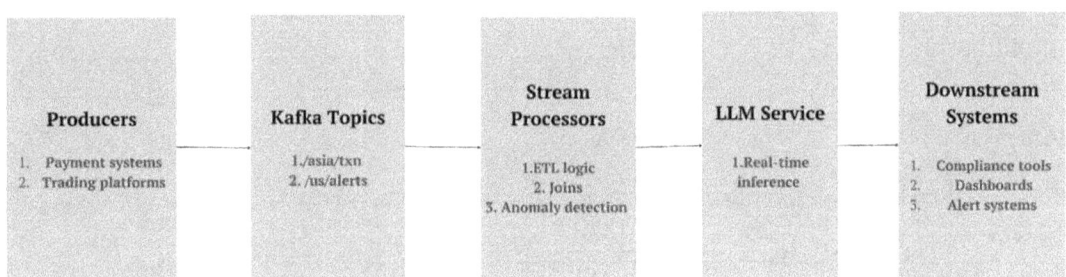

Figure 6-3. *Real-Time Streaming Pipeline Using Kafka*

Figure 6-3 outlines the flow of data from production systems to decision-making endpoints using Apache Kafka and large language models (LLMs). Below is a detailed explanation of each stage in the pipeline:

1. Producers

These are the **source systems** generating real-time data:

- **Payment Systems**: Generate transaction data—credit card swipes, digital wallet payments, ATM withdrawals, etc.

- **Trading Platforms**: Emit trading events—buy/sell orders, price updates, and portfolio movements.

These systems are often geographically distributed and produce a high volume of events that need to be processed immediately.

2. Kafka Topics

Apache Kafka serves as the **message broker** and streaming backbone:

- **/asia/txn**: Topic dedicated to transactions originating in Asia.

- **/us/alerts**: Topic for real-time alert signals generated from U.S.-based operations.

Kafka ensures that messages are stored durably, partitioned for scalability, and streamed to consumers in real time. It **decouples producers from downstream systems**, enabling flexibility and resilience.

3. Stream Processors

This component handles **on-the-fly data processing** before LLM inference:

- **ETL Logic**: Extracts, transforms, and loads incoming data to ensure consistency and cleanliness (e.g., removing nulls, timestamp formatting).

- **Joins**: Combines transaction streams with metadata (e.g., customer profiles or historical fraud scores).

- **Anomaly Detection**: Flags transactions or events that deviate from expected patterns using statistical or rule-based logic.

Stream processors **prepare and enrich** the data, reducing unnecessary load on the LLM service.

4. LLM Service

This is where **real-time inference** happens:

- **Real-Time Inference**: The enriched data is passed to a fine-tuned large language model that
  - Evaluates risk levels.
  - Classifies intent or sentiment.
  - Generates summaries or explanations.

The LLM acts as an **intelligent layer** that adds interpretability and predictive capabilities to the raw data.

5. Downstream Systems

These are the **end-user systems** that consume LLM outputs for actionable insights:

- **Compliance Tools**: Use LLM outputs to ensure regulatory alignment (e.g., anti-money laundering alerts, KYC validation).
- **Dashboards**: Display real-time analytics to decision-makers such as analysts or risk managers.
- **Alert Systems**: Trigger notifications (e.g., high-risk transaction flagged or abnormal market trend detected).

Downstream systems allow **real-time action** based on intelligent inferences from the pipeline.

Managing Throughput for High-Demand Applications

In high-volume financial environments—such as trading platforms, payment networks, and credit scoring systems—throughput becomes a critical performance metric. As the demand for real-time processing and intelligent decision-making increases, systems must be capable of handling thousands of concurrent requests per second without compromising accuracy or latency.

This section explores techniques to optimize throughput when integrating large language models (LLMs) into these high-demand systems. It covers strategies like batching requests, prioritizing critical workloads, and implementing dynamic autoscaling to ensure your architecture remains resilient, responsive, and resource-efficient under load. These practices are essential to maintain service quality, reduce infrastructure costs, and meet the real-time expectations of financial users.

Batching Requests: Group multiple small requests into a single batch to reduce overhead and improve efficiency. This approach works best for **non-time-critical workloads** where latency is less of a concern.

Caution Excessive batching can introduce delays that negatively impact **real-time latency SLAs**, especially in applications requiring instant responses.

Example: A compliance reporting system batches transactions from multiple accounts for simultaneous LLM processing.

Prioritizing Critical Requests: Assign higher priority to time-sensitive requests using a priority queue.

Example: A real-time trading system prioritizes price alert queries over less urgent customer interactions.

Monitoring and Autoscaling: Use monitoring tools to track metrics like request rates and latency. Enable autoscaling to add resources dynamically based on demand.

Example: A credit scoring system automatically scales up during loan application peaks.

Example: Real-Time Fraud Detection System

A bank implements an LLM-powered fraud detection system using real-time data processing techniques:

Architecture: Payment transactions are streamed to Apache Kafka. Kafka delivers data to an LLM-based fraud detection service hosted as a microservice.

Processing: The LLM analyzes each transaction in real time for anomalies. Suspicious transactions are flagged and sent to human investigators for review.

Outcome: The system achieves sub-second latency, preventing over $1 million in potential fraud losses annually.

Real-time data processing is an important aspect of integrating LLMs into financial systems. By using streaming platforms, optimizing for low latency, and adopting scalable architectures, financial institutions can harness the full potential of LLMs for time-sensitive applications. Techniques such as event-driven architectures, microservices, and edge computing ensure that LLMs operate efficiently in high-demand environments. These strategies not only enhance performance but also enable financial organizations to deliver faster, more accurate insights, ultimately driving better decision-making and operational success.

Case Studies

CASE STUDIES: SUCCESSFUL LLM INTEGRATIONS IN FINANCIAL SYSTEMS

Real-world examples of large language model (LLM) integrations in financial systems showcase the transformative potential of these models while highlighting practical approaches to overcome challenges. This section provides detailed case studies demonstrating how financial institutions have effectively implemented LLMs for customer service, market analysis, and compliance reporting. Each case study outlines the problem, solution, and outcomes, offering actionable insights for similar projects.

CASE STUDY 1: LLM-POWERED CUSTOMER SERVICE CHATBOTS

Problem

A large retail bank faced significant challenges in managing customer service operations due to a growing customer base and increasing call volumes. The existing system, reliant on human customer service representatives, struggled to meet the demand, leading to:

Caution:

While LLMs can handle large volumes of customer queries efficiently, they are also prone to **hallucinations,** plausible-sounding but incorrect responses. It is essential to implement **guardrails** that detect and correct such errors, such as confidence scoring, factuality checks, or **fallback-to-human review mechanisms** to maintain trust and compliance in customer interactions.

Long Wait Times:

- Customers experienced delays in getting their queries addressed, especially during peak hours.
- High-priority issues were often delayed due to the overwhelming number of routine queries.

Repetitive Queries:

- A significant portion of customer interactions involved repetitive questions, such as inquiries about account balances, credit card statements, and loan details.
- These queries consumed the time of service representatives, reducing their availability for complex issues.

Customer Dissatisfaction:

- Long wait times and inconsistent responses led to frustration and declining satisfaction scores.
- The bank received frequent complaints about poor service quality and delays.

High Operational Costs:

- The cost of maintaining a large team of customer service representatives was unsustainable.
- High attrition rates among representatives added to training and recruitment costs.

To address these challenges, the bank needed a scalable, efficient, and cost-effective solution to enhance customer service.

As shown in Table 6-2, LLMs in finance support use cases like fraud detection, customer service, and compliance reporting. Each role is enabled through API integrations such as fraud-check, chatbot, and compliance-report.

Table 6-2. LLM Use Cases in Finance and APIs

Use Case	LLM Role	API Integration Example
Fraud Detection	Classify transactions	/fraud-check
Customer Service	Respond to queries	/chatbot
Compliance Reporting	Generate reports	/compliance-report

Solution

The bank implemented an LLM-powered chatbot to streamline customer service operations and improve the overall customer experience. The solution involved integrating the chatbot into the bank's customer support system with the following components:

API Integration

The chatbot was deployed as a RESTful API, enabling seamless interaction with the bank's existing customer support platform.

Features of the API:

- Real-time query processing.
- Multi-channel support, including web, mobile apps, and messaging platforms like WhatsApp.

The API ensured consistent communication between the chatbot and the bank's backend systems, such as transaction databases and loan processing platforms.

Data Preprocessing and LLM Fine-Tuning

Historical chat logs and customer queries were cleaned and preprocessed to create a high-quality training dataset. Domain-specific data was used to fine-tune the LLM, enabling it to understand financial terminology and context.

Examples of fine-tuned queries:

- "What is my current account balance?"
- "How can I increase my credit card limit?"
- "What are the interest rates for personal loans?"

The LLM was trained to provide concise, accurate, and context-aware responses, improving its relevance and usability.

Note Fine-tuning is a cost-efficiency lever, and tailored models often outperform generic LLMs with significantly lower compute and inference costs.

Response Optimization

The chatbot was designed to handle a wide range of common queries, such as

- Checking account balances.
- Providing loan information.
- Resolving credit card payment issues.

CHAPTER 6   INTEGRATING LLMS INTO FINANCIAL SYSTEMS

An escalation mechanism was implemented to transfer unresolved or complex issues to human agents seamlessly.

Example: If the chatbot couldn't process a query like "I need assistance with a disputed transaction," it redirected the customer to a representative with the full context of the conversation.

Caution   Watch for bias in training data, as the financial LLMs trained on biased data may produce discriminatory or unfair outputs.

Personalized Interactions

The chatbot was integrated with the bank's CRM system to personalize responses based on the customer's profile and transaction history.

Example: When a customer asked about loan eligibility, the chatbot provided tailored recommendations based on their financial history.

Outcome

The deployment of the LLM-powered chatbot delivered significant improvements in customer service efficiency, satisfaction, and cost savings:

Increased Automation

- The chatbot handled over 60% of customer queries autonomously, significantly reducing the workload for human representatives.

- Common inquiries, such as account balances and transaction statuses, were resolved instantly, freeing up human agents to focus on complex cases.

Reduced Response Times

- The average response time decreased by 40%, as the chatbot provided instant replies to most queries.

- Customers no longer had to wait in queues for basic assistance, improving the overall service experience.

Improved Customer Satisfaction

- Customer satisfaction scores increased by 25%, driven by faster and more consistent responses.

- The chatbot's availability 24/7 ensured customers could get assistance at any time, further enhancing their experience.

Cost Savings

- The bank saved an estimated $2 million annually in operational costs by reducing dependency on human representatives and lowering recruitment and training expenses.

- Attrition rates among service representatives decreased, as they were able to focus on more engaging and value-added tasks.

Key Features of the Chatbot System

1. **Multi-Channel Support:** Accessible via web, mobile apps, and messaging platforms, ensuring convenience for customers.

2. **Secure Authentication:** Customers authenticated themselves using OTPs or biometric verification before accessing sensitive account information.

3. **Analytics and Feedback Loop:** The system tracked customer interactions and feedback to continuously improve the chatbot's performance through retraining.

Tip Incorporate feedback loops to include feedback from users and systems to retrain and improve LLMs periodically, ensuring relevance over time.

Lessons Learned

1. **Fine-Tuning Matters:** Tailoring the LLM to understand financial terminology and domain-specific queries was critical for achieving high accuracy and customer satisfaction.

2. **Escalation Mechanisms Are Essential:** The seamless handover of unresolved queries to human agents ensured a smooth customer experience and prevented dissatisfaction.

3. **Personalization Enhances Engagement:** Integrating the chatbot with customer profiles added a layer of personalization, making interactions more relevant and impactful.

4. **Multi-Channel Accessibility Drives Adoption:** Providing support across multiple platforms increased the chatbot's reach and usability.

CHAPTER 6   INTEGRATING LLMS INTO FINANCIAL SYSTEMS

The LLM-powered chatbot transformed the bank's customer service operations by addressing key pain points such as long wait times, high costs, and repetitive queries. By automating over half of the queries and providing instant, accurate responses, the chatbot not only improved operational efficiency but also enhanced customer satisfaction. This case study highlights the potential of LLMs to revolutionize customer support in the financial sector, providing a scalable and cost-effective solution for modern banking challenges.

CASE STUDY 2: REAL-TIME MARKET SENTIMENT ANALYSIS

An investment firm faced the challenge of analyzing real-time market sentiment from diverse data sources, including financial news, blogs, and social media platforms. Timely and accurate sentiment analysis was critical for providing actionable insights to traders, enabling them to make informed decisions in volatile markets. However, the firm encountered several obstacles:

Volume of Unstructured Data:

- News and social media platforms generated vast amounts of text data in real time, making manual analysis impractical.
- The data included diverse formats, writing styles, and potential noise, such as irrelevant or misleading content.

Need for Low Latency:

- Market conditions changed rapidly, and delays in processing sentiment data could result in missed trading opportunities.
- The existing systems were unable to provide actionable insights within the required timeframe.

Lack of Domain-Specific Customization:

- Generic sentiment analysis models lacked the specificity needed to understand financial jargon and context, reducing the accuracy of insights.

The firm needed a robust solution to process and analyze unstructured data in real time, classify sentiment accurately, and deliver insights with minimal latency.

Solution

The investment firm implemented an LLM-based market sentiment analysis system tailored to meet the demands of real-time financial decision-making.

Streaming Data Integration

Apache Kafka as the Backbone: Apache Kafka was used to ingest live data streams from multiple sources, including financial news outlets, blogs, and social media platforms like Twitter. Kafka topics were configured to categorize data based on the source and relevance, ensuring that only useful information was processed.

Data Preprocessing: Incoming data was preprocessed to remove noise, such as duplicate content, irrelevant posts, and spam. Natural language preprocessing techniques were applied to clean the text and prepare it for sentiment analysis.

To improve the granularity of sentiment analysis, the firm incorporated **Named Entity Recognition (NER)** techniques. NER enabled the extraction of entities such as company names, stock tickers, executive names, and geopolitical terms from unstructured text. By linking sentiment to specific entities, the system could provide more **targeted insights,** for example, distinguishing between positive sentiment about the tech sector overall vs. a specific company like NVIDIA.

LLM Fine-Tuning

Training with Domain-Specific Data: The LLM was fine-tuned on a dataset of labeled sentiment data derived from historical financial news and social media content. The dataset included annotations for positive, negative, and neutral sentiments, as well as metadata like source reliability and time relevance.

Customizing for Financial Jargon: The model was further trained to recognize domain-specific terminology, such as "bullish," "bearish," and "market rally," to improve classification accuracy.

Evaluation and Testing: The fine-tuned model was rigorously tested on real-world data to ensure high precision and recall.

Real-Time Processing

Low-Latency Architecture: A microservices architecture was deployed to process data streams and deliver sentiment scores in milliseconds. Each microservice handled a specific task, such as data ingestion, preprocessing, LLM inference, and result aggregation.

Caching and Optimization: Frequently queried data, such as trending topics, was cached to reduce redundant computations and improve response times.

Integration with Trader Dashboards: The system was integrated with real-time trading dashboards, where sentiment scores and summaries were displayed alongside other market indicators.

Outcome: The implementation of the LLM-based market sentiment analysis system resulted in substantial improvements in decision-making, efficiency, and portfolio performance:

Enhanced Trader Decision-Making:

- Traders gained access to sentiment scores and summaries in real time, enabling quicker responses to market shifts.
- For example, during a major economic event, the system flagged negative sentiment trends from social media, allowing traders to adjust positions before the broader market reacted.

Improved Portfolio Performance:

- By identifying sentiment-driven trends earlier than competitors, the firm improved portfolio returns by 15%.
- The system provided actionable insights that helped mitigate risks and capitalize on opportunities, such as sector-specific sentiment spikes.

Reduced Manual Effort:

- The system automated 80% of the effort required for analyzing market sentiment, freeing up analysts to focus on high-value tasks like strategy development and deep market analysis.
- Analysts previously spent hours sifting through news and social media; now, they relied on the system's outputs for initial insights.

Scalability and Reliability:

- The architecture supported scalability, allowing the system to handle surges in data volume during major market events without performance degradation.
- The firm could easily expand the system to include new data sources or integrate additional LLMs for advanced analytics.

CHAPTER 6   INTEGRATING LLMS INTO FINANCIAL SYSTEMS

Key Features of the System

1. **Sentiment Scoring and Summarization:** Each data point was classified as positive, negative, or neutral, with a confidence score. Summarized insights provided a concise overview of market sentiment, highlighting key drivers and trends.

2. **Custom Alerts:** Traders could set alerts for specific sentiment thresholds or keywords, such as "crisis" or "merger," ensuring they never missed critical information.

3. **Visualization Tools:** Sentiment trends were visualized in the trading dashboard, showing how sentiment evolved over time and correlating it with market movements.

Lessons Learned

Integrating LLMs into financial systems brings both technical and operational insights. This section highlights key lessons gained from real-world implementations, focusing on what truly drives performance and business value.

1. **Importance of Domain-Specific Fine-Tuning:** Tailoring the LLM for financial language and context was essential to achieving high accuracy and relevance in sentiment analysis.

2. **Value of Real-Time Processing:** The combination of streaming data integration and low-latency architecture ensured the system met the demands of time-sensitive trading decisions.

3. **Scalability for High-Volume Events:** Using Kafka and microservices provided the scalability needed to handle sudden increases in data, such as during earnings announcements or geopolitical events.

The LLM-based market sentiment analysis system transformed how the investment firm processed and utilized real-time data. By integrating streaming platforms, fine-tuning the LLM, and deploying a low-latency architecture, the firm gained a competitive edge in financial markets. This case study demonstrates the power of LLMs to deliver actionable insights in dynamic environments, offering a roadmap for other organizations seeking to harness the potential of AI for real-time decision-making.

CHAPTER 6   INTEGRATING LLMS INTO FINANCIAL SYSTEMS

CASE STUDY 3: AUTOMATED COMPLIANCE REPORTING

A multinational financial institution faced significant challenges in generating compliance reports to meet regulatory requirements. The institution operated across multiple jurisdictions, each with its own set of legal frameworks and reporting obligations. Compliance reports required a detailed analysis of vast volumes of transactional data, which included customer information, transaction histories, and audit trails.

The manual process of generating these reports involved

- **High Labor Intensity:** Teams of compliance officers manually reviewed transaction data and cross-referenced it with legal requirements.

- **Time Consumption:** Reports often took several days to complete, creating delays in meeting regulatory deadlines.

- **Error Risks:** The manual approach was prone to errors, such as missing critical compliance issues, which posed the risk of regulatory penalties and reputational damage.

As regulatory oversight increased, the institution needed an efficient, accurate, and scalable solution to handle compliance reporting.

Solution

To address these challenges, the institution developed an LLM-based compliance reporting system that automated the entire process. The system leveraged advanced natural language processing (NLP) capabilities of an LLM to analyze transactional data and generate structured reports aligned with legal frameworks.

Note   Regulatory compliance is non-negotiable, so ensure your integrations align with local and global financial regulations like GDPR, PCI-DSS, and MAS TRM.

To ensure transparency and trust in automated compliance systems, the institution adopted **traceability and explainability techniques**. These included maintaining **model logs**, generating **attention heatmaps** to visualize which data points influenced decisions, and integrating **decision tree overlays** for audit-friendly rule tracing. This allowed internal auditors and regulators to review how each compliance decision was derived—supporting accountability in multi-jurisdictional contexts.

Data Processing Pipeline: A data ingestion layer was built to stream transactional data from various systems, including payment gateways, customer management systems, and accounting platforms. The data underwent preprocessing to clean, normalize, and structure it for analysis. Key steps included

- Removing duplicates and handling missing data.
- Converting transaction data into formats compatible with the LLM's input requirements.
- Annotating data with relevant metadata, such as timestamps and jurisdiction codes.

LLM Integration: The LLM was fine-tuned using a curated dataset of legal documents, compliance frameworks, and historical reports. This ensured the model understood domain-specific language and requirements. The LLM was trained to

- Extract relevant information from transactional data.
- Identify compliance risks, such as suspicious transactions or breaches of regulatory thresholds.
- Generate structured reports with actionable insights, including risk summaries, flagged transactions, and recommendations.

API Deployment: The compliance system was exposed as a RESTful API, enabling compliance officers to interact with the system through a simple interface. Features of the API included

- **On-Demand Report Generation:** Officers could submit queries and receive comprehensive reports within minutes.
- **Customizable Outputs:** Reports could be tailored based on specific regulatory requirements or jurisdictions.
- **Secure Access:** The API included token-based authentication and encryption to ensure data security.

Outcome: The implementation of the LLM-based compliance reporting system led to significant improvements in efficiency, accuracy, and cost savings:

CHAPTER 6 INTEGRATING LLMS INTO FINANCIAL SYSTEMS

> **Caution** Avoid hard dependencies on specific APIs, as changing APIs can break your system—always design with abstraction and versioning.

Reduced Report Generation Time

- The time required to generate compliance reports decreased from several days to just a few hours.
- This improvement allowed the institution to meet tight regulatory deadlines and allocate resources more effectively.

High Accuracy

- The system achieved 98% accuracy in identifying compliance issues, significantly reducing errors and omissions in reports.
- By accurately flagging potential risks, the institution minimized the likelihood of regulatory penalties and enhanced its compliance posture.

Cost Savings

- Automation reduced the need for extensive manual effort, leading to annual savings of approximately $3 million in compliance-related operational costs.

Scalability

- The system could handle increased workloads during peak reporting periods without degradation in performance, ensuring consistency and reliability.
- Caution Secure every integration point, as the APIs and data pipelines are attack vectors—always enforce strong encryption and access control.

Improved Compliance Confidence

- Regulatory auditors provided positive feedback on the structured and detailed reports generated by the system, further strengthening the institution's reputation.

The adoption of an LLM-based compliance reporting system transformed how the institution managed regulatory obligations. By automating data analysis and report generation, the institution achieved faster, more accurate, and cost-effective compliance processes. This case study underscores the potential of LLMs to streamline complex workflows, improve operational efficiency, and enhance compliance in the highly regulated financial sector.

Financial institutions looking to modernize their compliance functions can draw valuable lessons from this implementation, particularly the integration of advanced NLP models with robust data pipelines and user-friendly APIs.

Caution Beware of model drift as the financial data changes over time; unmonitored LLMs can quickly become obsolete.

Key Takeaways from Case Studies

1. **API Integration:** Seamless integration through APIs allowed LLMs to connect with existing financial systems efficiently.

2. **Domain-Specific Fine-Tuning:** Tailoring LLMs to specific financial use cases improved their accuracy and relevance.

3. **Real-Time Processing:** Low-latency architectures ensured that the models could meet the demands of time-sensitive financial applications.

4. **Scalable Solutions:** By leveraging scalable infrastructure and tools like Kafka, institutions handled high volumes of data without performance degradation.

5. **Cost Savings and Efficiency Gains:** Automation and LLM integration reduced manual effort, improved accuracy, and generated significant cost savings.

These case studies illustrate the practical challenges and solutions involved in integrating LLMs into financial systems. Whether it's enhancing customer service, analyzing market sentiment, or automating compliance reporting, LLMs have proven to be transformative tools. By adopting similar strategies—such as API-based integration, real-time processing, and domain-specific fine-tuning—financial institutions can unlock the full potential of LLMs to drive innovation, efficiency, and customer satisfaction in their operations.

Conclusion

Integrating large language models (LLMs) into financial systems is a pivotal step toward leveraging artificial intelligence for enhanced operational efficiency, decision-making, and customer satisfaction. This chapter outlined the essential strategies and best practices for achieving seamless integration, emphasizing the importance of scalable API development, real-time data processing, and domain-specific customization.

By addressing challenges like system compatibility and data security, financial institutions can build robust and reliable systems that maximize the potential of LLMs. The real-world case studies presented highlight the transformative impact of well-executed integrations, offering actionable insights for similar projects.

While integration lays the foundation for operationalizing LLMs, ensuring their long-term performance requires robust monitoring and maintenance practices. The next chapter will explore how to track critical performance metrics, detect and address model drift, and implement automated retraining workflows. You will learn strategies to keep their LLMs accurate, reliable, and compliant over time, preparing them for sustained success in dynamic financial environments.

Points to Remember

1. **APIs Are the Backbone of Integration:** APIs enable seamless interaction between LLMs and financial systems, ensuring efficient communication and data exchange.

2. **Focus on Scalable API Design:** Use RESTful APIs with features like versioning, backward compatibility, and security protocols to support long-term system stability.

3. **Real-Time Data Processing Is Crucial:** Streaming platforms like Apache Kafka facilitate low-latency data processing, enabling real-time applications such as fraud detection and market analysis.

4. **Scalability Ensures Adaptability:** Implement horizontal scaling, load balancing, and microservices architectures to handle growing workloads effectively.

5. **Domain-Specific Fine-Tuning Adds Value:** Customizing LLMs for specific financial tasks improves accuracy and relevance, enhancing their utility.

6. **Address Security and Compliance:** Use encryption, authentication, and adherence to data privacy regulations to safeguard sensitive financial information.

7. **Leverage Automation for Efficiency:** Automate workflows for data ingestion, preprocessing, and retraining to streamline operations and reduce manual effort.

8. **Adopt Edge Computing for Low Latency:** Deploying LLMs on edge devices can reduce dependency on centralized systems and improve response times.

9. **Monitor Performance Continuously:** Keep track of system metrics like latency, throughput, and accuracy to identify areas for improvement and ensure operational stability.

10. **Learn from Real-World Examples:** Case studies demonstrate the effectiveness of LLMs in improving customer service, market analysis, and compliance reporting, offering practical insights for implementation.

CHAPTER 7

Monitoring and Maintenance of LLMs

Large language models (LLMs) are at the core of many financial applications, where accuracy, reliability, and adaptability are essential for success. However, deploying an LLM is just the beginning—ensuring its long-term effectiveness requires continuous monitoring and maintenance. Over time, performance degradation, data drift, and evolving user needs can impact model accuracy and relevance. If left unchecked, these issues can lead to financial losses, compliance risks, and reduced trust in AI-driven decisions. This chapter explores key strategies for monitoring LLM performance, the importance of regular updates, and best practices for handling model drift. By implementing robust maintenance workflows, organizations can proactively detect performance issues, retrain models with fresh data, and optimize their LLMs for long-term efficiency. With a structured approach to LLM monitoring and maintenance, financial institutions can extend model lifespan, enhance decision-making accuracy, and ensure compliance with evolving regulatory standards.

Structure

This chapter covers the following topics:

- **Introduction to Monitoring and Maintenance**: Why continuous monitoring is essential for accuracy, adaptability, and compliance.
- **Performance Metrics**: Tracking accuracy, latency, resource utilization, and fraud detection precision with real-time monitoring tools.

- **Regular Updates and Retraining**: Implementing incremental retraining, active learning, and automated pipelines to adapt to evolving data.

- **Handling Model Drift**: Detecting and mitigating data drift and concept drift through statistical analysis and adaptive learning.

- **Challenges and Best Practices**: Addressing computational costs, data privacy, and scalability with alert thresholds, A/B testing, and workflow optimization.

- **Future Trends**: Exploring predictive maintenance, self-healing AI, and reinforcement learning for next-generation LLM operations.

Objectives

By the end of this chapter, you will understand how to ensure that large language models (LLMs) remain reliable, accurate, and efficient. Organizations must implement robust monitoring and maintenance strategies. This chapter provides the knowledge and tools needed to track performance, detect issues, and optimize LLM operations over time.

You will learn how to monitor key performance metrics such as accuracy, latency, and resource utilization using real-time monitoring frameworks. The chapter also explores strategies for regular updates and retraining, including incremental learning, active learning, and automated retraining pipelines, ensuring that LLMs adapt to evolving data and business needs.

Additionally, this chapter covers handling model drift, explaining its causes and offering detection and mitigation techniques to prevent performance degradation. You will also discover best practices for scalable maintenance workflows, minimizing downtime, and ensuring data security.

Finally, the chapter introduces emerging trends, such as predictive maintenance and self-healing AI systems, preparing you for the next generation of LLM operations. By mastering these concepts, you will be able to maintain high-performing, resilient, and adaptive LLMs, ensuring their seamless integration into financial decision-making and business processes.

Introduction

The deployment of large language models (LLMs) in finance has transformed key areas such as fraud detection, customer service, credit scoring, and regulatory compliance. However, deployment is just the beginning—ensuring long-term performance requires continuous monitoring and maintenance. Without a structured approach, models may degrade over time, fail to adapt to changing data patterns, or violate regulatory standards, leading to financial and reputational risks.

This section explores why monitoring and maintaining LLMs is essential for their continued success in financial applications. It highlights common challenges in dynamic environments and provides actionable strategies to address them effectively.

Ensuring Long-Term Performance

Financial LLMs operate in high-stakes environments, where accuracy and reliability are critical. Performance degradation, whether due to data shifts or operational inefficiencies, can result in significant financial losses.

- **Fraud Detection Risks**: A model that fails to detect new fraud patterns exposes financial institutions to monetary and compliance risks.

- **Credit Scoring Errors**: A credit scoring model that does not adapt to changing consumer behavior can lead to incorrect creditworthiness assessments, affecting both borrowers and lenders.

Regular monitoring and maintenance prevent performance degradation, ensuring that LLMs continue to deliver reliable results.

Adapting to Evolving Data

Financial data is highly dynamic, meaning the statistical properties of market trends, consumer behaviors, and transaction types can change over time. LLMs must adapt to these shifts to remain effective.

- **Data Drift**: A shift in financial transactions, such as an increase in digital payments over cash transactions, can impact model performance if not accounted for. It is the change in the statistical properties of the input data over time (e.g., changes in feature distributions such as the mean or variance). For instance, an increase in digital payments over cash transactions can shift the data distribution, affecting model performance if not accounted for.

- **Concept Drift**: Changes in relationships between variables, such as new fraud patterns during economic downturns, require models to be updated frequently. Occurs when the underlying relationship between input features and the target variable changes. This can happen when fraud patterns evolve or user behavior shifts. Statistically, this implies a change in the conditional distribution $P(y|X)$ over time, even if $P(X)$ remains stable.

  Popular concept drift detection methods include

- **ADWIN (Adaptive Windowing):** Dynamically detects change by adjusting the size of a sliding window.

- **DDM (Drift Detection Method):** Monitors prediction error rate and signals drift when it increases beyond a threshold.

Monitoring systems must detect these changes early, triggering retraining workflows to keep LLMs aligned with real-world trends.

As shown in Table 7-1, retraining strategies include incremental retraining, full retraining, and active learning. Each balances efficiency, coverage, and focus, with use cases from chatbot updates to fraud detection.

Table 7-1. Retraining Strategies

Strategy	Advantages	Example Use Case
Incremental Retraining	Efficient, keeps prior knowledge	Chatbot vocabulary updates
Full Retraining	Comprehensive, includes all data	New fraud pattern inclusion
Active Learning	Focuses on the most informative data	Uncertain sentiment classification

Compliance with Operational Requirements

The financial sector is heavily regulated, requiring AI models to comply with data privacy, fairness, and transparency guidelines. Continuous monitoring ensures

- **Bias Detection**: Identifying unfair outcomes in credit scoring, fraud detection, or loan approvals to ensure compliance with fairness regulations.

- **Decision Explainability**: Maintaining audit trails and model transparency, which are essential for regulatory reviews and risk management.

Failing to monitor compliance risks regulatory penalties, reputational damage, and loss of customer trust.

Challenges in Monitoring and Maintenance

Monitoring and maintaining LLMs in finance comes with unique challenges, including

- **Volume and Velocity of Data**: Financial institutions process massive amounts of real-time data, requiring efficient monitoring solutions to track performance continuously.

- **Computational Costs**: Frequent model retraining demands significant computational resources, making cost-effective strategies essential.

- **Integration with Legacy Systems**: Many financial organizations rely on older infrastructure, complicating model monitoring and deployment workflows.

Overcoming these challenges requires automated, scalable monitoring and maintenance practices.

Examples of Monitoring

Successful monitoring and maintenance strategies ensure that LLMs remain accurate, compliant, and high-performing in financial applications. The following case studies illustrate how financial institutions have leveraged real-time tracking, automated updates, and optimization techniques to sustain LLM reliability and efficiency.

Real-Time Fraud Detection

A financial institution deployed an LLM for fraud detection but noticed a decline in accuracy as fraud tactics evolved. Real-time monitoring flagged the issue, triggering automated retraining with new transaction data. As a result, the model regained accuracy and prevented financial losses.

Compliance in Credit Scoring

A credit scoring firm monitored its LLM-based credit risk model and detected bias against certain demographic groups. By fine-tuning the training data and model parameters, the company ensured compliance with fair lending regulations, avoiding potential legal penalties.

Market Sentiment Analysis

An asset management firm used an LLM to analyze financial news and social media sentiment. During peak trading hours, monitoring systems identified latency issues affecting decision-making speed. By optimizing the inference pipeline, the firm ensured timely, accurate insights for traders.

Monitoring and maintenance are not just operational tasks—they are critical to the long-term success of LLMs in finance. By implementing robust tracking, updating models proactively, and ensuring regulatory compliance, financial institutions can safeguard performance, reduce risks, and maximize the value of AI-driven decision-making.

In this chapter you will learn the strategies and tools needed to establish scalable, efficient monitoring and maintenance workflows, ensuring that LLMs remain reliable, adaptive, and compliant in complex financial environments.

Performance Metrics

Monitoring the performance of large language models (LLMs) is essential to ensuring their reliability, efficiency, and effectiveness in financial applications. Well-defined performance metrics help organizations assess how models are functioning and identify areas for optimization and retraining. This section explores key performance metrics, their relevance to finance, and real-time monitoring tools that enable proactive management of LLM operations.

As shown in Table 7-2, key performance metrics for LLMs include accuracy, latency, resource utilization, precision, and sentiment analysis accuracy. These metrics ensure reliable monitoring across use cases like fraud detection, trading, and customer feedback.

Table 7-2. Key Performance Metrics

Metric	Definition	Example Use Case
Accuracy	Correct predictions over total predictions	Fraud detection model
Latency	Time taken to generate output	Algorithmic trading model
Resource Utilization	CPU, GPU, memory usage efficiency	Credit scoring system
Fraud Detection Precision	True positives / (True positives + False positives)	Transaction monitoring
Sentiment Analysis Accuracy	Correct sentiment classification	Customer feedback analysis

Key Performance Metrics for LLMs

When evaluating the performance of large language models (LLMs) in financial applications, it is essential to consider several key metrics that provide insights into how well the models are functioning. Below are some of the most important metrics, including their definitions, examples in finance, best practices, and concluding insights.

Accuracy

Accuracy is a fundamental metric that measures the proportion of correct predictions made by a model. In financial settings, ensuring high accuracy is vital because any misclassification can result in financial losses, regulatory penalties, and a loss of trust from customers.

Example in Finance

For example, in fraud detection, accuracy refers to the percentage of transactions that are correctly classified as fraudulent or legitimate. A model with low accuracy could either wrongly flag legitimate transactions as fraudulent or miss actual fraudulent transactions, both of which could have costly consequences.

Best Practices

It's essential to track other metrics like precision (positive predictive value) and recall (true positive rate) alongside accuracy for a more balanced evaluation. This provides a fuller picture of how the model is performing, especially in cases where the data is imbalanced, such as with fraud detection.

Accuracy is an important metric, but it is not sufficient by itself. A comprehensive evaluation using precision and recall is crucial for mitigating financial and operational risks in financial applications.

Recall
Recall measures the proportion of actual positive cases that were correctly identified by the model. In financial use cases such as fraud detection or anti-money laundering (AML), recall is especially critical; missing true positives (i.e., failing to catch fraudulent transactions) can lead to severe losses.

F1-Score
The F1-score is the harmonic mean of precision and recall, offering a balanced measure when there is an uneven class distribution. This is particularly useful in finance, where false positives (e.g., flagging a legitimate transaction as fraudulent) and false negatives both carry high costs.

Area Under the ROC Curve (AUC-ROC)
AUC-ROC evaluates the model's ability to distinguish between classes across various threshold settings. It's widely used for binary classification tasks and is essential when the cost of misclassification varies, as in credit scoring or compliance violation predictions.

Latency

Latency refers to the time it takes for a model to generate an output after receiving input. In high-stakes financial applications where real-time decisions are necessary, low latency is critical for timely actions. However, there's often a trade-off between **latency and accuracy**. Models optimized for high accuracy (e.g., large, deep architectures) may require more computational time, resulting in increased latency. Conversely, ultra-low-latency models may sacrifice some accuracy for faster inference—a compromise that must be carefully managed based on the use case.

Example in Finance
In algorithmic trading, a model with high latency could miss a window of opportunity for executing trades. Even a slight delay in making decisions could result in missed profits or poor trade execution, especially in fast-moving markets.

Best Practices
To maintain low latency, optimize the model's inference pipelines. This can be achieved by using techniques such as caching or precomputing frequently queried data to ensure faster response times without compromising the accuracy of predictions.

Minimizing latency is essential for ensuring that financial systems can respond in real time, which is critical for applications such as high-frequency trading, fraud detection, and risk management.

Resource Utilization

Resource utilization refers to how efficiently a model uses computational resources such as CPU, GPU, memory, and storage. Efficient resource management is key to maintaining cost-effectiveness and scalability, particularly when operating at scale.

Example in Finance
Consider a credit scoring model running in resource-constrained environments. Balancing performance with resource consumption ensures that the system remains operational and cost-effective, especially in high-volume environments where frequent model updates are required.

Best Practices
Monitor resource utilization trends to detect potential bottlenecks that could hinder performance. Additionally, model compression techniques such as pruning or quantization can be employed to reduce the model's computational demands without sacrificing accuracy or functionality.

Optimizing resource utilization is vital for ensuring that financial models are scalable, cost-efficient, and capable of handling increasing volumes of data and transactions.

Domain-Specific Metrics

Certain financial applications require specialized metrics to evaluate the model's success in fulfilling its unique role. These metrics provide deeper insights into how well the LLM meets specific industry needs.

Fraud Detection Precision
Precision in fraud detection measures the percentage of flagged transactions that are actually fraudulent. A high precision score reduces false positives, which minimizes the need for costly manual investigations and improves overall efficiency.

Risk Assessment Accuracy

In credit scoring, loan approvals, and market analysis, risk assessment accuracy measures the effectiveness of a model in predicting the financial risk of clients or investments. High accuracy in this area ensures that financial decisions are based on reliable predictions, reducing potential losses.

Customer Sentiment Analysis

For customer feedback and financial news sentiment analysis, this metric evaluates how accurately the model can classify the sentiment expressed in text. Accurate sentiment analysis allows institutions to make more informed decisions regarding customer relations and market strategies.

Domain-specific metrics are essential for ensuring that LLMs in finance deliver the expected outcomes. By aligning model performance with industry-specific needs, organizations can enhance operational efficiency and decision-making.

Real-Time Metric Monitoring

Real-time monitoring is crucial for tracking the performance of LLMs and identifying potential issues before they affect operations. This ensures that models remain effective, responsive, and aligned with business objectives. Below are some essential monitoring tools and their use cases in the financial sector.

Prometheus and Grafana

Prometheus is an open-source system that collects and stores time-series data, making it ideal for real-time performance monitoring. Grafana complements Prometheus by visualizing this data through customizable dashboards, enabling easy interpretation and analysis.

Use Case in Finance

In a trading system powered by an LLM, Prometheus and Grafana can be used to monitor critical metrics such as latency, throughput, and resource utilization. By integrating both tools, financial institutions can gain real-time visibility into the performance of their LLMs, enabling the proactive identification and resolution of issues before they affect operations.

Combining Prometheus and Grafana provides comprehensive real-time monitoring, ensuring smooth operation and timely intervention in high-stakes financial environments.

Evidently AI

Evidently AI is a performance monitoring tool designed for machine learning models. It tracks key metrics like precision, recall, and accuracy and can detect data drift over time, ensuring that the model continues to perform optimally.

Use Case in Finance
Evidently AI is particularly useful in fraud detection systems, where it can track the model's performance over time. By identifying any potential drift early, financial institutions can retrain models or adjust thresholds to maintain high accuracy and reduce the risk of missed fraud detection.

Evidently AI allows financial organizations to keep track of model performance and ensure timely adjustments, minimizing the risk of inaccuracies in critical systems like fraud detection.

Google Vertex AI Monitoring

Google Vertex AI provides integrated monitoring for models deployed on Google Cloud, with advanced anomaly detection capabilities. This tool allows for real-time tracking of various performance metrics and can detect irregularities that may indicate underlying issues.

Use Case in Finance
For customer service chatbots powered by LLMs, Google Vertex AI can monitor latency and accuracy. By identifying any deviations from expected performance, it helps ensure that the chatbot remains responsive and accurate, maintaining high-quality customer interactions.

Google Vertex AI enhances the stability of cloud-based LLMs by providing real-time anomaly detection, helping to maintain a seamless user experience and efficient operations.

AWS CloudWatch

AWS CloudWatch tracks the resource utilization and system performance of applications running on AWS infrastructure. It provides detailed metrics and can send alerts when predefined thresholds are breached, helping to manage resources proactively.

Use Case in Finance

In a financial institution, CloudWatch can be used to ensure that LLMs running on AWS are optimized for resource usage. By setting up alerts for critical metrics, such as CPU or memory usage, organizations can quickly address issues and maintain system performance without overburdening infrastructure.

AWS CloudWatch ensures that financial institutions can maintain optimal resource utilization and performance by providing timely alerts and actionable insights.

Custom Monitoring Pipelines

For organizations with specific needs, custom monitoring pipelines can be created using tools like Apache Kafka for real-time data streaming and Elasticsearch for advanced analysis. This enables the creation of tailored monitoring solutions that meet the unique demands of a business.

Use Case in Finance

Custom monitoring pipelines can be used to track sentiment analysis models during critical market events. By streaming real-time data through Kafka and analyzing it with Elasticsearch, financial institutions can assess model performance in dynamic environments and make adjustments as necessary.

Custom monitoring solutions offer flexibility and precision, allowing financial institutions to optimize LLM performance in rapidly changing market conditions and other dynamic situations.

By using these real-time monitoring tools, financial institutions can ensure that their LLMs are continuously performing at their best, minimizing risk, optimizing performance, and maintaining efficiency across various applications.

Best Practices for Performance Monitoring

Effective performance monitoring is essential for ensuring that LLMs remain stable, accurate, and efficient in financial applications. Below are best practices that can help organizations establish robust monitoring systems to maintain high-performing models.

Define Thresholds for Key Metrics

Establishing clear thresholds for key performance metrics such as accuracy, latency, and resource utilization is crucial for ensuring optimal model performance. By setting acceptable ranges for these metrics, organizations can quickly identify when the model deviates from expected performance levels.

Example
For real-time trading models, set a maximum latency threshold of 200ms to ensure swift decision-making and execution.

Defining thresholds provides clear performance benchmarks, allowing organizations to take timely corrective actions when necessary.

Implement Alerts

Using monitoring tools to send automated alerts when performance deviates from predefined thresholds is essential for proactive issue resolution. Alerts provide early warnings, allowing teams to address problems before they escalate and negatively impact operations.

Example
Set up an alert to trigger if the accuracy of a fraud detection model drops below 95%. This will prompt the team to investigate the cause of the decline and take corrective action.

Alerts act as an early warning system, ensuring that financial institutions can respond swiftly to any issues and minimize risks.

Regularly Review Metrics

Periodic reviews of performance metrics are critical for identifying long-term trends and uncovering opportunities for optimization. This practice helps organizations stay ahead of potential issues, ensure continuous improvement, and adjust strategies as needed.

Example
Schedule quarterly reviews to assess whether the fraud detection model's accuracy is maintaining or improving over time and adjust training data or model parameters accordingly.

Regular metric reviews provide valuable insights into model performance, helping to refine strategies and keep LLMs aligned with business goals.

CHAPTER 7 MONITORING AND MAINTENANCE OF LLMS

Correlate Metrics for Deeper Insights

Analyzing the relationships between different performance metrics provides deeper insights into the model's overall behavior. For example, understanding the correlation between resource utilization, latency, and accuracy can help organizations optimize for balanced performance, improving efficiency while maintaining accuracy.

Example

Examine how increased resource consumption (e.g., memory usage) impacts latency in a real-time trading system. If high resource usage results in slower response times, adjustments can be made to improve the balance between resources and model performance.

Correlating metrics helps uncover hidden patterns that can guide optimization efforts, leading to more efficient and effective LLMs in financial applications.

Defining and tracking performance metrics is a foundational aspect of effective LLM monitoring and maintenance. Key metrics such as accuracy, latency, and resource utilization provide essential insights into model performance, while domain-specific measures enhance financial applications. Real-time monitoring tools play a critical role in detecting potential issues early, allowing for proactive troubleshooting and retraining. By adopting best practices like defining thresholds, setting up alerts, regularly reviewing metrics, and correlating data, financial institutions can maintain high-performing LLMs that continuously adapt to evolving data and business needs.

A structured approach to performance monitoring is crucial for achieving scalability, compliance, and reliability in financial AI applications. Implementing these best practices ensures that LLMs remain stable, accurate, and cost-efficient in a dynamic and complex industry.

Tip Regularly benchmark model performance using historical data comparisons to detect subtle performance drifts before they cause major issues.

To ensure stability in financial AI systems, institutions should define **acceptable deviation thresholds** for key metrics. For instance, a drop of more than **2% in fraud recall** might trigger model retraining or rollback protocols.

Additionally, adopting **SLA/SLO frameworks,** such as expecting **95% of LLM API responses within 300ms,** helps maintain performance consistency across latency-sensitive services.

To proactively detect issues, organizations can apply **anomaly detection** methods like **Facebook Prophet**, **Isolation Forest**, or rolling z-score techniques to identify abnormal metric trends over time.

> **Note** Monitoring is not just about accuracy—focusing on latency, resource efficiency, and domain-specific metrics ensures a balanced and optimized LLM deployment.

Regular Updates and Retraining

Large language models (LLMs) are dynamic systems that require continuous updates and retraining to stay accurate, relevant, and compliant in financial applications. As financial markets, fraud strategies, and consumer behaviors constantly evolve, LLMs must adapt to these changes. Without regular updates, LLMs may provide outdated predictions, leading to poor decision-making, compliance issues, and financial risks. This section explores the importance of regular updates, effective retraining strategies, and the role of automation in maintaining strong LLM performance.

As shown in Table 7-3, model drift can take the form of data drift, concept drift, or combined drift. In finance, this is seen in shifts like changing payment methods, evolving fraud tactics, or simultaneous market and sentiment changes.

Table 7-3. Types of Model Drift

Drift Type	Description	Example in Finance
Data Drift	Change in input data distribution	Rise in digital vs. cash payments
Concept Drift	Change in relationship between features and outcome	New fraud tactics in economic downturns
Combined Drift	Simultaneous data and concept drift	Sentiment and market impact change together

Adapting to Changing Data

Financial data is ever-changing, shaped by economic shifts, market fluctuations, and emerging fraud tactics. If LLMs are only trained on historical data, they may lose accuracy as new patterns emerge.

CHAPTER 7   MONITORING AND MAINTENANCE OF LLMS

Example

A credit scoring model trained on data from before the COVID-19 pandemic might struggle to evaluate risks effectively in a post-pandemic economic environment.

Solution

Regular updates ensure that LLMs stay aligned with current data trends, improving both prediction accuracy and reliability. Continuous retraining based on new data allows models to adapt to market and customer behavior changes, maintaining their effectiveness over time.

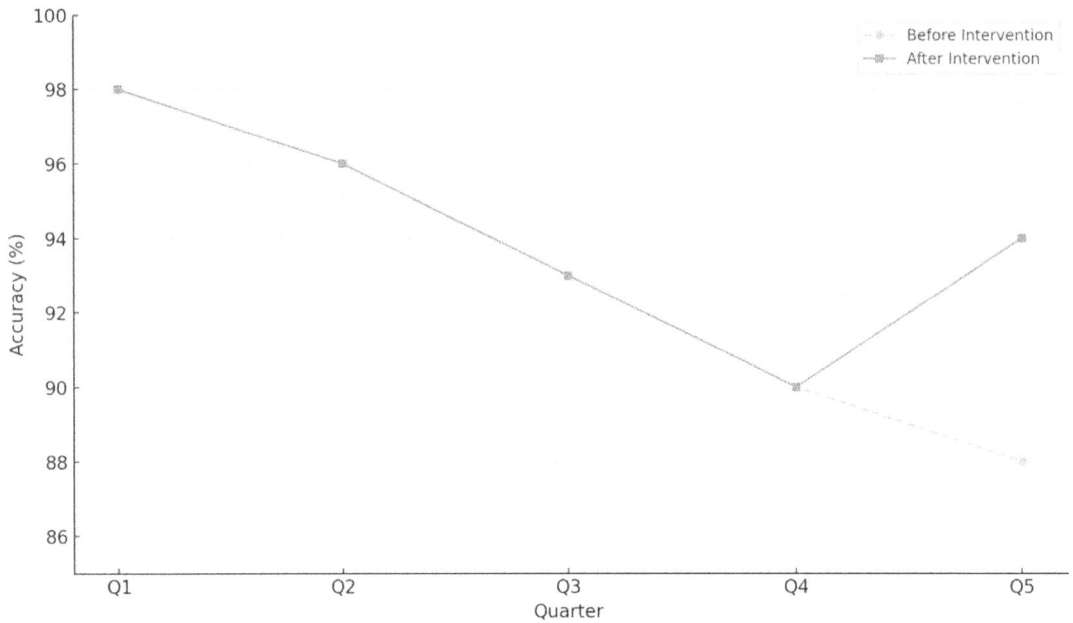

Figure 7-1. *Model Accuracy Over Time*

Figure 7-1 illustrates the performance trajectory of a financial LLM across five quarters. Initially, the model exhibits a steady decline in accuracy—from 98% in Q1 to 88% in Q4—indicating potential data or concept drift. Following targeted intervention, such as retraining with recent data or tuning hyperparameters, accuracy improves significantly to 94% in Q5. This visual underscores the importance of continuous monitoring and proactive maintenance. Timely interventions can restore model reliability and prevent the compounding effects of performance degradation in high-stakes financial applications.

Responding to Regulatory Changes

Financial regulations are continually evolving, making it essential for LLMs to stay updated to comply with new standards. Failure to update LLMs can lead to legal risks and compliance violations.

Example
Changes in anti-money laundering (AML) laws might require fraud detection models to incorporate new transaction monitoring patterns.

Solution
Routine model updates ensure that financial AI applications stay compliant with evolving regulations, mitigating legal risks while enhancing transparency in financial operations.

Addressing Model Drift

Over time, models experience drift, where their predictions become less accurate due to shifts in data or underlying relationships between variables.

- **Data Drift**: Changes in the distribution of input data (e.g., the rise of digital payments replacing cash transactions).

- **Concept Drift**: Shifts in the relationships between input variables and output results (e.g., the emergence of new fraud techniques).

Solution
Regular retraining addresses drift, ensuring that models continue to provide relevant, accurate insights in rapidly changing financial environments.

Strategies for Retraining LLMs

There are several approaches to retraining LLMs, each suited for different needs and circumstances. The following strategies help maintain the performance and relevance of models while balancing computational efficiency.

Incremental Retraining

Incremental retraining involves updating the model with only the latest data, allowing it to retain prior knowledge while incorporating new trends and patterns.

Advantages

- Efficient from a computational perspective, as it avoids the need for full retraining.
- Ideal for handling small, frequent updates that reflect gradual shifts in data.

Example in Finance

Updating a customer service chatbot with new customer queries and updated financial terminology.

Benefit

Incremental retraining enables continuous model evolution with minimal computational overhead.

Full Retraining

Full retraining involves rebuilding the model from scratch using the entire updated dataset, incorporating both new and historical data.

Advantages

- Comprehensive and ensures all changes, both old and new, are integrated.
- Eliminates the risk of errors compounding over time due to incremental updates.

Challenges

- High computational costs and longer retraining times.

Example in Finance

Rebuilding a fraud detection model to capture entirely new transaction patterns that arise due to evolving fraud tactics.

Benefit

Full retraining ensures models remain robust and accurate but requires significant computational resources.

Active Learning

Active learning focuses on prioritizing the most informative data points for retraining. The model selects the data it needs the most to improve its performance, allowing for more efficient updates.

Advantages

- Reduces unnecessary retraining by selecting the most relevant data.
- Enhances learning efficiency and speeds up model improvements.

Example in Finance

Using active learning to retrain a sentiment analysis model with customer feedback that falls near the decision boundary, improving its classification accuracy.

Benefit

Active learning ensures that models are trained only on the most useful data, enhancing retraining efficiency and performance.

Regular updates and retraining are essential for maintaining the accuracy and relevance of LLMs in financial applications. By adapting to changing data, responding to regulatory changes, and addressing model drift, organizations can ensure that their models remain effective in dynamic environments. Strategies like incremental retraining, full retraining, and active learning provide flexibility in how models are updated, enabling financial institutions to balance computational costs and model performance effectively.

CHAPTER 7   MONITORING AND MAINTENANCE OF LLMS

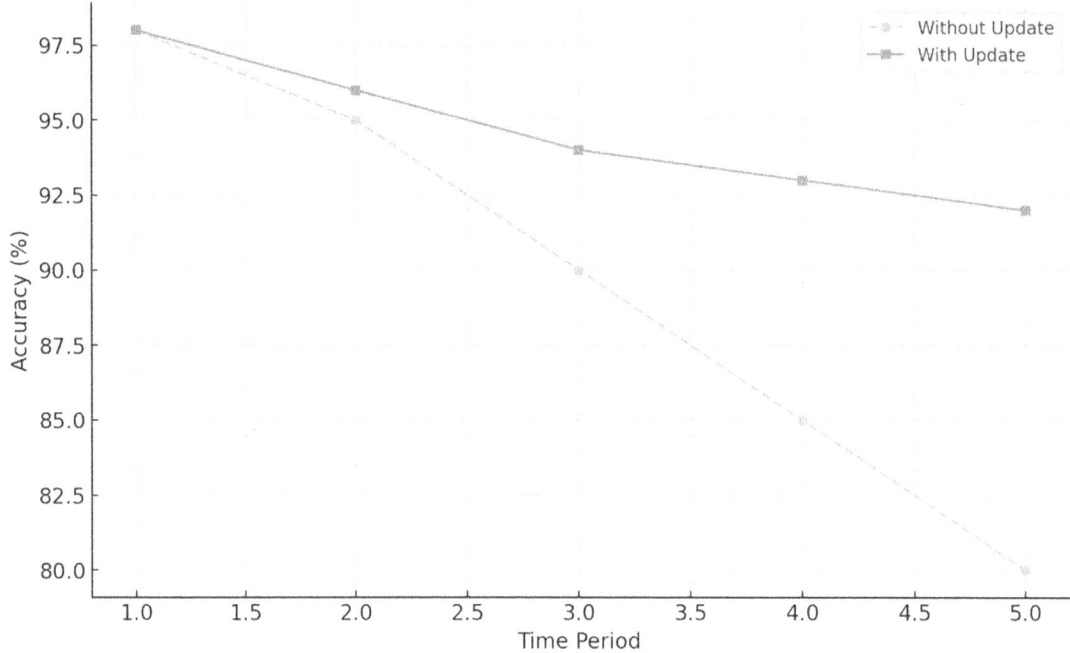

Figure 7-2. *Effect of Drift on Model Accuracy*

This comparative line plot shows how model accuracy degrades over time due to drift if not addressed (from 98% to 80%). In contrast, models that undergo periodic updates maintain higher accuracy levels (from 98% to 92%). Figure 7-2 underscores the necessity of active drift detection and retraining strategies to preserve model performance in dynamic environments.

Automating Retraining Pipelines

Automation plays a critical role in optimizing retraining workflows, reducing manual intervention, and ensuring timely model updates. By automating key aspects of model updates, organizations can improve efficiency, minimize errors, and ensure that models stay up-to-date with evolving data. This results in more reliable predictions and better performance over time.

Components of an Automated Pipeline

An automated retraining pipeline includes several essential components. First, **data collection** ensures that new data from various sources—such as transactions, market trends, and customer interactions—are automatically ingested. This keeps the model continuously updated with fresh information. The next step is **data cleaning and preprocessing**, which removes noise, corrects biases, and standardizes input formats. This step ensures that the data being fed into the model is of the highest quality and ready for use. Once the data is preprocessed, **model training and validation** are automated, allowing for consistent benchmarking against previous versions to ensure the new model's reliability. Finally, the **deployment** phase replaces outdated models with updated versions in production, enabling the system to continue operating without disruptions.

Benefits of Automation

The benefits of automation in the retraining process are significant. **Efficiency** is one of the primary advantages, as automating repetitive tasks eliminates bottlenecks and saves time and resources. Additionally, automation ensures **consistency** by following standardized workflows, which minimizes the risk of human error during retraining. Moreover, automation allows for greater **scalability**, as financial institutions can deploy and manage large-scale LLM systems across multiple applications without compromising performance or reliability.

Example of an Automated Retraining Pipeline

To illustrate the benefits of automation, consider a bank that implemented an automated retraining pipeline for its fraud detection model. The system collects new transaction data daily from multiple sources, preprocesses it, and retrains the model with incremental updates. Once retrained, the new model is validated using test data to ensure it performs as expected. Finally, the updated model is automatically deployed into production. This automated process has led to a 20% reduction in fraud detection errors, demonstrating the effectiveness of automation in improving model accuracy and operational efficiency.

Therefore, automating the retraining pipeline ensures that LLMs remain accurate, scalable, and adaptable, which is essential for maintaining high performance in financial applications.

CHAPTER 7   MONITORING AND MAINTENANCE OF LLMS

Best Practices for Regular Updates and Retraining

Regular updates and retraining are essential for maintaining the accuracy, relevance, and compliance of large language models (LLMs) in financial applications. Best practices for ensuring effective retraining include defining an appropriate update frequency based on data volatility, such as setting weekly or monthly retraining schedules. Trigger-based retraining can be employed, where models are automatically retrained if performance falls below certain thresholds, like an accuracy drop below 95%. It's also crucial to monitor resource usage to optimize infrastructure and balance cost with scalability. Ensuring data quality through audits for biases and inconsistencies helps maintain model reliability, while human oversight is important for periodically reviewing retrained models to ensure they align with business goals and regulatory requirements. By incorporating strategies such as incremental retraining, full retraining, and active learning, organizations can keep LLMs aligned with real-time data and evolving regulations. Automation of retraining pipelines further enhances efficiency and scalability, allowing for continuous model improvement without manual intervention. Adopting these systematic retraining practices helps financial institutions prevent model degradation, improve decision-making accuracy, and ensure regulatory compliance, enabling LLMs to deliver reliable and high-quality outcomes in an ever-changing financial landscape.

Tip   Regularly benchmark model performance using historical data comparisons to detect subtle performance drifts before they cause major issues. Implement adaptive retraining schedules—frequent incremental updates for fast-changing applications and periodic full retraining for comprehensive improvements.

Note   Automated retraining does not eliminate the need for human oversight. Periodic model audits ensure updates align with business goals and compliance standards.

Caution Ignoring retraining can cause performance degradation, increased bias, and regulatory non-compliance. Regular updates are essential to maintaining model effectiveness.

Handling Model Drift

Model drift is one of the **most significant challenges** faced by **large language models (LLMs)** in real-world applications, particularly in **finance**, where **data patterns, regulatory landscapes, and economic conditions** change rapidly. Drift occurs when **shifts in data distributions or relationships between variables** lead to a **decline in model accuracy and reliability**. Understanding, detecting, and mitigating drift is **critical** to ensuring LLMs remain **effective, compliant, and aligned** with business objectives.

To address this, organizations can apply **time-decay retraining** strategies, where more recent data is weighted more heavily than older data. This helps models adapt to changing patterns without requiring full retraining.

Additionally, **reservoir sampling** is useful in online LLM adaptation, allowing models to maintain a representative and bounded sample of the most relevant incoming data over time, even in streaming environments.

Understanding, detecting, and mitigating drift is critical to ensuring LLMs remain effective, compliant, and aligned with business objectives.

This section explores **types of model drift**, their **impact on LLM performance**, **detection techniques**, and **strategies to mitigate drift** effectively.

Types of Model Drift

As large language models (LLMs) are deployed in real-world financial environments, their performance can degrade over time due to evolving data patterns. In this section you will learn the different types of model drift—data drift, concept drift, and combined drift—and their unique impacts on LLM performance.

Data Drift
Data drift occurs when the statistical properties of input data change over time, making the training data no longer representative of real-world conditions. This can significantly

affect the performance of LLMs, as the model's ability to generalize to new inputs is reduced, leading to incorrect predictions or misclassifications. In finance, for instance, a fraud detection model trained on pre-pandemic transaction data may struggle to identify fraud patterns in a post-pandemic world where online transactions have surged. Detecting and addressing data drift ensures that LLMs continue to reflect evolving market trends and customer behaviors, maintaining their relevance and accuracy.

Concept Drift

Concept drift happens when the relationship between input features and target variables changes over time. This causes the model to rely on outdated correlations, impacting predictions and leading to decreased accuracy. In finance, a credit scoring model may fail to adapt to changing economic conditions, where income levels and default rates no longer correlate as they did before an economic downturn. Understanding and detecting concept drift helps organizations ensure that their financial models remain predictive and reliable, even as the underlying conditions shift.

Combined Drift

Combined drift occurs when both data drift and concept drift happen simultaneously, complicating efforts to maintain model accuracy. This amplification of degradation requires comprehensive retraining and constant monitoring to ensure that the model performs well. For example, a market sentiment analysis model may experience combined drift when both the language used in financial news and the relationship between sentiment and stock performance change. To handle combined drift effectively, organizations need a multi-faceted approach, incorporating frequent retraining and adaptive learning techniques.

Caution Ignoring real-time performance monitoring can result in unexpected failures, leading to financial losses, compliance violations, and operational disruptions. Proactive tracking is essential to prevent these risks.

Detection Methods for Model Drift

Statistical Tests for Data Drift

Statistical tests provide an early warning system for detecting significant shifts in data distributions. Common techniques include the Kolmogorov-Smirnov test, which compares input data distributions from training and inference; Wasserstein distance, which measures differences between two probability distributions; and the chi-square test, which identifies changes in categorical data. For example, a bank might use the Kolmogorov-Smirnov test to detect shifts in transaction amount distributions, which helps adjust fraud detection thresholds. These statistical methods are essential for identifying data drift early on, enabling timely interventions.

Performance Monitoring for Concept Drift
To detect concept drift, it's crucial to regularly evaluate the model's performance metrics, such as accuracy, precision, recall, and F1-score. Tracking performance trends over time can highlight degradation, signaling that concept drift may be occurring. For example, a decline in a credit scoring model's F1-score might indicate that the model is no longer capturing the relationships between economic variables accurately, prompting further investigation. Regular performance monitoring is key to detecting concept drift before it significantly impacts decision-making, ensuring models stay effective.

Real-Time Drift Detection Tools
Real-time drift detection tools such as Evidently AI, Alibi Detect, and custom monitoring pipelines allow organizations to continuously monitor both data and concept drift in machine learning pipelines. These tools use statistical and machine learning techniques to detect drift as it happens, enabling quick responses to prevent business disruptions. For example, a real-time drift detection tool could alert a financial institution when a sudden shift in transaction data signals that the fraud detection model needs an update. These tools are critical in minimizing the impact of drift and ensuring that LLMs remain accurate and reliable.

Statistical Techniques for Drift Analysis

Feature Importance Analysis
Feature importance analysis tracks changes in the relevance of input features over time, revealing shifts in the relationships between features and the target variable. This technique helps to identify which features are becoming more or less influential in predicting outcomes, which is crucial for maintaining the model's relevance. In finance, for example, a sentiment analysis model might detect that the term "inflation" becomes less significant as market conditions stabilize. By tracking feature importance, organizations can identify emerging trends and optimize model predictions to better align with evolving data.

Covariate Shift Detection

Covariate shift detection focuses on identifying changes in the distribution of input features, without assuming that the target variable (or the model's predictions) has shifted. This method is particularly useful when the relationship between the inputs and outputs remains relatively stable, but the data itself evolves. In finance, for example, detecting a surge in digital transactions within a payment processing dataset could indicate a behavioral change among consumers. By using covariate shift detection, financial institutions can ensure that their models remain aligned with evolving data sources and adjust their predictions accordingly.

Predictive Monitoring

Predictive monitoring involves comparing a model's predictions against ground truth data in real-time to identify performance deviations. This technique is essential for catching any discrepancies between the model's outputs and actual outcomes, especially during periods of significant change. In finance, for instance, a trading model's predictions may diverge significantly from actual stock movements during volatile market conditions. Predictive monitoring provides real-time insights into how well the model is performing and highlights when corrections are needed to maintain accuracy.

Strategies for Mitigating Model Drift

Mitigating model drift is essential to maintain the accuracy and relevance of LLMs in dynamic financial environments. This section outlines practical strategies—such as regular monitoring, retraining, and adaptive learning—that help detect drift early and respond effectively.

Regular Monitoring and Alerts

Proactively monitoring model performance and setting performance thresholds can help identify drift early, triggering alerts when necessary. This strategy is essential for addressing issues before they significantly impact model accuracy. For example, an anomaly detection system might flag unusual transaction patterns in a payment processing system for further investigation. Regular monitoring and timely alerts ensure that financial institutions can detect drift early and take corrective actions to minimize disruptions.

Note Scalable LLM monitoring workflows reduce operational complexity and improve efficiency.

Incremental Retraining

Incremental retraining involves updating models with recent data to ensure they remain reflective of current trends. This approach is especially useful when drift is gradual and the changes in data are not drastic. For instance, a bank might retrain its customer service chatbot using recently logged queries to adapt to evolving customer interactions. Incremental retraining is a cost-efficient way to keep models current, as it allows for gradual updates without the need for a full model overhaul.

> **Tip** Use automated monitoring dashboards to visualize key performance indicators for quick anomaly detection.

Full Retraining

Full retraining is necessary when drift is significant and incremental updates alone are insufficient to maintain model accuracy. This strategy involves retraining the model from scratch using the most recent data, effectively resetting the model's knowledge base. For example, a credit scoring model may undergo full retraining after a major economic shift that alters the correlations between features and outcomes. Full retraining is critical for restoring accuracy after major disruptions, ensuring that the model remains aligned with the latest data and trends.

Adaptive Learning Techniques

Adaptive learning techniques, such as reinforcement learning or online learning, enable models to continuously adapt to new data without requiring frequent manual retraining. These techniques are particularly useful for real-time decision-making scenarios, where the model needs to adjust its predictions dynamically based on new information. For example, a portfolio management model might use reinforcement learning to adjust investment strategies in real-time based on market conditions. Adaptive learning ensures that LLMs remain relevant and effective without the need for constant retraining, making them more resilient to shifts in the data.

> **Tip** Combine reinforcement learning with self-healing workflows to continuously improve performance without frequent retraining.

Ensemble Models

Ensemble models combine multiple models to reduce the impact of drift and enhance overall performance. By integrating different approaches, such as rule-based systems and machine learning models, ensemble methods increase resilience against sudden data and concept shifts. For example, a fraud detection system might integrate a rule-based model with an LLM to provide a more robust and adaptable solution for identifying fraudulent activities. Ensemble models help ensure that the system remains accurate even when faced with complex or unexpected changes in the data.

Case Study: Handling Drift in Fraud Detection

A financial institution deployed a large language model (LLM) for fraud detection but noticed a decline in its accuracy due to evolving fraud patterns. In response, the company

- Implemented a drift detection pipeline using Evidently AI, which successfully identified shifts in transaction frequency.

- Conducted weekly incremental retraining to incorporate the latest fraud data and ensure the model remained up-to-date.

- Integrated a rule-based system alongside the LLM to handle unexpected changes in fraud schemes.

Outcome: False negatives decreased by 25%, significantly improving the accuracy of fraud detection.

By combining automated drift detection and adaptive retraining strategies, the institution ensured that its LLM continued to perform effectively in real-world financial scenarios.

Addressing model drift is an ongoing and critical aspect of maintaining LLMs in the finance sector. Understanding the different types of drift, applying effective detection methods, and utilizing mitigation strategies are key to preserving the accuracy and relevance of models. Proactively managing drift enhances operational efficiency while minimizing financial and reputational risks.

The techniques outlined in this section empower financial professionals and engineers to effectively address drift, ensuring that LLMs consistently deliver high-quality, reliable predictions in dynamic financial environments.

Challenges and Best Practices

Monitoring and maintaining large language models (LLMs) in finance requires careful planning to ensure they remain accurate, compliant, and efficient in dynamic data environments. Financial institutions use LLMs for critical tasks like fraud detection, credit risk analysis, and customer support. Continuous monitoring is essential to prevent performance degradation, ensure regulatory compliance, and optimize operations. However, these benefits come with challenges such as computational costs, data privacy concerns, scalability issues, and model drift.

This section discusses the challenges associated with managing LLMs and offers best practices for creating efficient, scalable monitoring workflows that enhance model reliability and compliance.

Challenges in Monitoring and Maintaining LLMs

LLMs require significant computational resources for training, inference, and retraining, often demanding high-speed GPUs and cloud services. These requirements can increase operational costs, especially in real-time financial applications.

Impact

- Elevated infrastructure costs, particularly with multiple models in production.
- Slower response times and inefficiencies during peak demand periods.

Example in Finance

A credit risk model used for instant loan approvals consumes substantial GPU resources, significantly escalating cloud computing costs during peak financial demand.

Solution

Optimizing computational efficiency is essential for balancing performance with cost-effectiveness.

Data Privacy and Security

Problem

Financial institutions handle sensitive customer data, which raises compliance concerns with privacy regulations like GDPR, CCPA, and industry-specific banking rules. Monitoring LLMs often involves processing personal financial data, necessitating robust data protection measures.

Impact

- Non-compliance with data privacy regulations can lead to legal penalties and loss of customer trust.
- Increased risks of security breaches and cyberattacks.

Example in Finance

Retraining a fraud detection model requires anonymized transaction data, which adds complexity and preprocessing overhead.

Solution

Implementing strong data privacy safeguards ensures compliance and protects customer data.

Scalability Challenges

As financial institutions deploy more LLMs for various functions, such as fraud detection, risk analysis, and customer service, managing multiple models becomes increasingly complex.

Impact

- Inconsistent monitoring workflows can cause performance discrepancies between models.
- Manual tracking becomes inefficient, leading to delays in addressing issues.

Example in Finance

A bank struggling to centralize monitoring across LLMs used for fraud detection, loan underwriting, and customer service experiences inefficiencies.

Solution

Adopting scalable model management practices ensures consistent monitoring and optimized performance across all applications.

Model Drift Detection and Mitigation

Data and concept drift can cause models to become outdated, undermining their ability to provide accurate financial predictions.

Impact

- Unaddressed drift results in incorrect decisions, affecting critical areas like risk management and fraud detection.
- Decreased model accuracy can jeopardize regulatory compliance.

Example in Finance

A trading model fails to adapt to new market conditions, leading to poor trade execution and financial losses.

Solution

Early drift detection and proactive retraining are essential to maintain model relevance and accuracy in rapidly changing markets.

Coordination Between Teams

Effective monitoring and maintenance of LLMs requires collaboration between data scientists, engineers, compliance officers, and financial analysts. Misaligned priorities and inefficient communication can slow down critical updates and hinder optimal performance.

Impact

- Delayed model updates due to misaligned team priorities.
- Potential compliance issues or suboptimal model performance.

Example in Finance

Engineering teams focus on optimizing model efficiency, while compliance teams prioritize regulatory requirements, causing delays in deploying necessary updates.

Solution

Clear communication and collaboration strategies between teams prevent operational bottlenecks and ensure timely model updates.

Best Practices for Monitoring and Maintenance

Effective monitoring and maintenance are key to ensuring that LLMs remain accurate, compliant, and efficient over time. In this section you will learn the best practices that help financial institutions build scalable, reliable, and proactive AI management workflows.

Define Alert Thresholds

What to Do?
Set predefined thresholds for accuracy, latency, and drift indicators. Configure automated alerts when models deviate beyond acceptable ranges.

Why It Matters?
Early detection of performance degradation allows for swift action to prevent downtime.

Example in Finance
An alert is triggered if the accuracy of a fraud detection model drops below 90% or if latency exceeds 200ms.

Solution
Real-time alerts help ensure consistent model performance and quick resolution of issues.

Implement A/B Testing

What to Do?
Deploy updated models alongside existing versions in a controlled environment. Compare performance metrics to ensure improvements before full deployment.

Why It Matters
Reduces risks by validating model updates and ensuring they enhance performance.

Example in Finance
A customer support chatbot undergoes A/B testing to evaluate improvements in response accuracy.

Solution
A/B testing guarantees that updates positively impact performance before full implementation.

Adopt Scalable Maintenance Workflows

What to Do?
Use centralized monitoring platforms for managing multiple models across applications. Leverage cloud-based solutions for scalability.

Why It Matters
Ensures consistent monitoring of all financial models, promoting operational efficiency.

Example in Finance
A financial institution uses Google Vertex AI to monitor and manage all deployed models from a single platform.

Solution
A centralized monitoring strategy streamlines model management and ensures consistency.

Automate Monitoring and Retraining Pipelines

What to Do?
Automate workflows for data ingestion, drift detection, retraining, and deployment.

Why It Matters
Automation reduces manual workload, accelerates updates, and minimizes model downtime.

Example in Finance
An automated pipeline updates a market sentiment analysis model daily with the latest news data.

Solution
Automating processes enhances efficiency and ensures continuous model optimization.

Ensure Data Privacy and Compliance

What to Do?
Implement data anonymization, encryption, and access control measures. Conduct regular compliance audits to ensure adherence to privacy regulations.

CHAPTER 7 MONITORING AND MAINTENANCE OF LLMS

Why It Matters

Protects customer data and helps institutions avoid legal risks associated with non-compliance.

Example in Finance

A bank uses differential privacy techniques to prevent data traceability during model training.

Solution

Compliance-focused monitoring builds customer trust and prevents legal complications.

Monitor Resource Utilization

What to Do?

Track CPU, GPU, and memory usage to optimize resource allocation and minimize waste.

Why It Matters

Balancing model performance with cost-efficiency helps optimize resources.

Example in Finance

A trading platform uses a quantized version of its LLM to reduce GPU resource costs.

Solution

Efficient resource utilization improves scalability and financial AI performance.

By following these best practices, financial institutions can effectively manage and maintain LLMs, ensuring they remain performant, compliant, and cost-efficient in a fast-paced, ever-changing environment.

Caution   Ignoring resource consumption can result in excessive costs—optimize model deployments strategically.

Future Trends

As financial institutions increasingly depend on large language models (LLMs) for key operations, traditional methods of model monitoring and maintenance are evolving. The next wave of LLM maintenance will leverage predictive maintenance, self-healing systems, and reinforcement learning-driven adaptability. These cutting-edge technologies will enable AI systems to autonomously detect, diagnose, and resolve performance issues, ensuring high efficiency, scalability, and resilience in financial

CHAPTER 7 MONITORING AND MAINTENANCE OF LLMS

applications. This section delves into these emerging trends, their potential applications, and how organizations can prepare for the future of LLM maintenance.

As shown in Table 7-4, future trends include predictive maintenance, self-healing systems, reinforcement learning, edge computing, and blockchain integration. These innovations aim to enhance resilience, adaptability, and security in AI systems.

Table 7-4. Future Trends

Trend	Description
Predictive Maintenance	Forecast issues before failure
Self-Healing Systems	Auto-diagnose and correct errors
Reinforcement Learning	Learn continuously from environment
Edge Computing	On-device inference and updates
Blockchain Integration	Secure update tracking and audit

Caution   Failing to integrate self-healing and predictive maintenance could result in costly downtimes, inaccurate predictions, and compliance risks—adopt these trends early to remain competitive.

Predictive Maintenance

How It Works

Predictive maintenance uses real-time data and machine learning algorithms to forecast potential failures before they affect model performance. By analyzing accuracy trends, drift indicators, and resource consumption, predictive systems can schedule retraining and updates only when necessary, preventing disruptions and optimizing costs.

Benefits

- **Minimized Downtime**: Detects issues early, ensuring continuous operation.

- **Cost Efficiency**: Reduces unnecessary retraining by focusing resources on models at risk.

- **Improved Performance**: Keeps LLMs relevant and aligned with evolving financial data trends.

Example in Finance
A fraud detection system monitors changes in transaction patterns. When anomalies suggest data drift, predictive maintenance triggers incremental retraining, avoiding declines in fraud detection accuracy.

Future Applications

- **Edge Computing Integration**: Real-time, on-device analysis for predictive maintenance at the edge.

- **Cross-Model Predictive Systems**: Centralized systems managing multiple LLMs across various applications (e.g., fraud prevention, risk management, and compliance).

Self-Healing Systems

How It Works
Self-healing systems allow LLMs to autonomously detect, diagnose, and resolve performance issues. These systems combine

- **Automated monitoring** for continuous evaluation.
- **Error correction mechanisms** that adjust parameters in real-time.
- **Targeted retraining pipelines** to fine-tune specific aspects of the model.

Benefits

- **Autonomy**: Reduces manual maintenance efforts, enabling self-adjustment in dynamic environments.

- **Scalability**: Supports large-scale deployments with interacting models.

- **Consistency**: Ensures models stay aligned with evolving financial conditions.

Example in Finance
A market sentiment analysis model detects new terminology during an economic downturn. The self-healing system collects updated data, fine-tunes the model, and redeploys an optimized version within hours.

Future Applications

- **Blockchain Integration**: Secure, auditable logs of model updates stored on blockchain.

- **Autonomous Hyperparameter Tuning**: AI-driven optimization for tasks like credit risk assessment and portfolio management.

Reinforcement Learning-Driven Adaptability

How It Works
Reinforcement learning (RL) enables LLMs to adapt continuously by interacting with their environment and learning from feedback. Models receive rewards or penalties based on their performance, refining decision-making strategies without explicit retraining.

Benefits

- **Dynamic Adaptation**: Allows LLMs to adjust in real time to evolving financial conditions.

- **Reduced Retraining**: Continuous learning minimizes the need for scheduled retraining cycles.

- **Enhanced Performance**: Improves long-term accuracy in applications like trading and credit risk assessment.

Example in Finance
A trading model adapts its investment strategies based on real-time market conditions, optimizing performance without manual updates.

Future Applications

- **RL-Powered Customer Support**: Chatbots that refine responses based on user interactions for better financial advisory services.

- **Adaptive Credit Scoring Models**: LLMs that adjust risk assessment criteria as market conditions and borrower behaviors change.

CHAPTER 7   MONITORING AND MAINTENANCE OF LLMS

The Intersection of Trends

Predictive maintenance, self-healing systems, and reinforcement learning are interconnected and can complement each other to create a robust AI ecosystem for financial institutions.

- Predictive maintenance can trigger self-healing systems when anomalies are detected.
- Self-healing models can use reinforcement learning to optimize their own corrections.

By integrating these strategies, financial institutions can develop self-sufficient, intelligent, and adaptive AI systems that are resilient, scalable, and efficient.

Preparing for the Future of LLM Maintenance

As shown in Table 7-5, monitoring tools range from Prometheus + Grafana for metrics to Evidently AI for drift detection, and cloud-native options like Vertex AI and CloudWatch. Custom pipelines enable tailored monitoring with Kafka and Elasticsearch.

Table 7-5. Monitoring Tools

Tool	Functionality
Prometheus + Grafana	Time-series metric collection and dashboards
Evidently AI	Model performance and drift detection
Google Vertex AI	Cloud model monitoring and anomaly alerts
AWS CloudWatch	Infrastructure usage tracking and alerts
Custom Pipelines	Tailored monitoring using Kafka and Elasticsearch

Invest in Advanced Monitoring Tools

- **What to Do**: Adopt AI-driven monitoring tools that offer real-time predictive analysis and automated anomaly detection.
- **Why It Matters**: Proactive maintenance minimizes downtime and failures, ensuring continuous service.

Build Scalable Infrastructure

- **What to Do**: Ensure that your infrastructure supports self-healing workflows and reinforcement learning-driven optimization.
- **Why It Matters**: Scalable infrastructure helps prevent bottlenecks, enabling seamless model expansion.

Focus on Interdisciplinary Collaboration

- **What to Do**: Foster collaboration between data scientists, engineers, and financial experts to ensure alignment with business goals.
- **Why It Matters**: Collaboration ensures that technical solutions align with business priorities, driving more effective AI outcomes.

Embrace Continuous Learning

- **What to Do**: Stay informed about emerging AI advancements and best practices.
- **Why It Matters**: Keeping up with new technologies ensures that institutions remain at the forefront of LLM maintenance strategies.

The future of LLM maintenance lies in the development of autonomous, proactive, and intelligent systems. Predictive maintenance, self-healing models, and reinforcement learning-driven adaptability will reshape how financial institutions manage LLMs, ensuring long-term sustainability, resilience, and competitive advantage. By embracing these trends, organizations can unlock the full potential of AI-driven finance, keeping pace with the evolving demands of the industry.

> **Note** The shift from reactive to proactive LLM maintenance will future-proof financial AI operations. Implement predictive systems now to stay ahead.

Conclusion

Maintaining large language models (LLMs) in finance is no longer just about monitoring performance and retraining models periodically. As AI-driven financial applications grow in complexity, the future of LLM maintenance demands proactive, autonomous, and adaptive systems. Predictive maintenance, self-healing systems, and reinforcement learning-driven adaptability represent groundbreaking advancements that will transform how financial institutions manage AI models, ensuring they remain resilient, efficient, and continuously aligned with evolving market dynamics.

By implementing predictive analytics, organizations can detect performance issues before they occur, minimizing downtime and optimizing retraining schedules. Self-healing systems introduce real-time model correction, reducing dependency on manual interventions. Meanwhile, reinforcement learning-based adaptability allows LLMs to continuously refine their decision-making capabilities without frequent retraining, improving long-term model sustainability.

Together, these emerging trends pave the way for a next-generation LLM ecosystem—one that is scalable, intelligent, and self-sufficient. Financial institutions that embrace these innovations early will not only enhance model reliability and regulatory compliance but also gain a competitive edge in AI-driven finance.

As we move forward, the next chapter explores the critical topic of scaling LLMs in financial applications. Managing model deployments across multiple financial services, optimizing infrastructure, and ensuring efficient resource utilization are key to maintaining high-performance AI systems at scale. Join us as we dive into the best practices for scaling LLMs while maintaining robustness and compliance in high-stakes financial environments.

Key Points

1. **LLM Maintenance Is Shifting Toward Proactive Approaches**: Traditional monitoring and retraining are being replaced by predictive maintenance and self-healing systems, enabling early detection and resolution of issues.

2. **Predictive Maintenance Prevents Performance Degradation**: By analyzing real-time data trends, predictive systems can identify potential failures before they happen, reducing downtime and unnecessary retraining.

3. **Self-Healing Systems Enable Autonomous Model Adjustments**: These systems use automated monitoring, error correction, and retraining pipelines to ensure LLMs stay aligned with real-world data without human intervention.

4. **Reinforcement Learning-Driven Adaptability Enhances Long-Term Model Performance**: LLMs continuously refine their decision-making by interacting with real-time financial data and receiving feedback, reducing the need for frequent manual retraining.

5. **Emerging Trends Work Together to Create Resilient AI Systems**: Predictive maintenance, self-healing AI, and reinforcement learning are complementary technologies that collectively improve the efficiency, scalability, and robustness of LLMs.

6. **Financial Institutions Must Invest in Advanced Monitoring Tools**: AI-driven real-time anomaly detection and performance tracking ensure financial LLMs remain stable and compliant with evolving regulations.

7. **Scalable Infrastructure Is Essential for Future LLM Operations**: Financial organizations must build flexible, cloud-based architectures that support self-healing workflows and adaptive learning mechanisms.

8. **Interdisciplinary Collaboration Is Key**: Successful LLM maintenance requires seamless coordination between data scientists, engineers, compliance teams, and financial analysts to align AI strategies with business objectives.

9. **Automating Maintenance Reduces Costs and Enhances Efficiency**: Automated pipelines for data ingestion, drift detection, and model retraining streamline operations, lowering computational costs while improving performance.

10. **The Future of LLM Maintenance Lies in Fully Autonomous Systems**: By integrating predictive analytics, self-healing models, and reinforcement learning, financial institutions can create AI-driven ecosystems that are self-sufficient, scalable, and continuously improving.

CHAPTER 8

Future Trends in LLM Ops for Finance

The landscape of large language models (LLMs) in finance is evolving rapidly, driven by technological breakthroughs and shifting regulatory landscapes. In this chapter, we explore the future of LLM operations, providing insights into how advancements in quantum computing, model architectures, and predictive maintenance are set to redefine financial AI applications.

Beyond technology, we discuss the growing importance of regulatory compliance and ethical AI in shaping financial AI strategies. As new standards emerge, financial institutions must proactively adapt to maintain a competitive edge.

You'll also learn about predictive maintenance—a critical approach that enables real-time monitoring and automated model optimization, ensuring AI systems remain resilient in dynamic market conditions.

By staying ahead of these trends, you will gain a strategic advantage in the next generation of AI-driven finance.

Structure

This chapter covers the following topics:

- **Overview of Evolving LLM in Finance**: Explore the evolving role of LLMs in finance.
- **Emerging Technologies:** Exploration of quantum computing and advancements in LLM architectures that could revolutionize model operations.

- **Predictive Maintenance of Models:** Techniques for proactive monitoring and automating updates based on predictive insights, ensuring long-term model relevance and efficiency.

- **The Future of AI in Finance:** Insights into expected shifts in regulations, ethical standards, and the emergence of new financial applications for LLMs.

Objectives

By the end of this chapter, you will have a forward-looking perspective on the key trends shaping the future of large language models (LLMs) in finance.

You'll explore emerging technologies such as quantum computing and next-generation model architectures, which promise to enhance computational efficiency and unlock new AI capabilities. You'll also learn about predictive maintenance strategies, including proactive monitoring and automated updates, to keep models optimized in rapidly changing financial environments.

Beyond technology, this chapter examines the evolving regulatory landscape and ethical considerations that will influence the future of AI-driven finance. You'll gain insights into new applications of LLMs, helping you anticipate industry shifts and adapt your strategies accordingly.

Armed with this knowledge, you will be well-prepared to navigate the next wave of AI innovations in finance, ensuring your models remain efficient, compliant, and ahead of the curve.

Overview of Evolving LLMs in Finance

The financial sector is undergoing significant transformations as it integrates advanced technologies like large language models (LLMs) into its processes. These models have become central to tasks such as credit scoring, fraud detection, customer support, and market analysis. However, as the applications of LLMs expand, so too do the challenges associated with managing them effectively. This chapter sets the stage by examining the evolving field of LLMs in finance, the importance of staying responsive to technological and regulatory developments, and the topics that will be explored in detail.

The use of LLMs in finance has grown significantly, driven by their ability to process vast amounts of data, identify patterns, and provide actionable insights. From enhancing risk management to improving customer service experiences, these models have become a vital tool for financial institutions aiming to stay competitive.

Below, I explore key factors driving the evolution of LLMs in finance:

- **Growth in Data Complexity**

    Financial institutions generate and handle large volumes of unstructured data, including transaction histories, customer communications, and regulatory reports. Increasingly, this data includes complex formats such as time-series data from market feeds and graph data representing relationships among entities. LLMs provide the capability to extract meaningful insights from this data, enabling more informed decision-making.

- **Expanding Use Cases**

    Initially applied to customer support and document analysis, LLMs now support complex applications such as scenario analysis, real-time fraud detection, and compliance automation. Their ability to adapt to various tasks makes them a key component in modern finance. Table 8-1 summarizes various use cases of large language models (LLMs) in the financial industry, mapping each application area to the specific role played by LLMs.

Table 8-1. Financial Applications in LLMs

Application	LLM Role
Fraud Detection	Detect patterns and anomalies in transactions
Credit Risk Scoring	Generate predictive credit risk scores
Customer Support	Automate query handling with natural language
Compliance Automation	Parse and analyze legal documents
Market Sentiment Analysis	Interpret social/news data for investment insights

CHAPTER 8   FUTURE TRENDS IN LLM OPS FOR FINANCE

Adapting to Regulatory Changes

As LLMs become deeply embedded in financial systems, staying ahead of technological advancements and regulatory shifts is essential. Ignoring these changes can lead to inefficiencies, increased risks, and missed opportunities. To successfully integrate LLMs in finance, institutions must adapt to rapid technological advancements, evolving regulations, and ethical considerations. Below are the key factors shaping this dynamic landscape:

Keeping Up with Technological Advancements

The rapid pace of AI innovation means that models and infrastructure can quickly become outdated. To maintain efficiency and relevance, financial institutions must continuously assess and update their LLM operations. Advancements in quantum computing, edge devices, and optimized architectures are reshaping how LLMs are deployed and scaled, making it crucial to stay agile and adopt cutting-edge solutions.

Navigating Evolving Regulatory Frameworks

Regulatory standards for financial AI are becoming increasingly stringent, with a focus on transparency, explainability, and data privacy. Institutions must ensure compliance with key regulations such as GDPR and Basel III, which emphasize data security, risk mitigation, and ethical AI practices. Adhering to these frameworks is critical to avoiding legal and reputational risks.

Addressing Ethical Considerations

Beyond compliance, financial institutions must tackle ethical challenges such as bias in AI outputs, fairness in decision-making, and responsible model usage. Developing robust strategies for accountability and fairness is essential for building trust with regulators, customers, and stakeholders. A proactive approach to ethical AI ensures long-term sustainability and credibility in the evolving financial landscape.

By staying proactive in adopting new technologies, ensuring regulatory compliance, and addressing ethical challenges, financial institutions can maximize the potential of LLMs while minimizing risks. A well-structured approach to LLM operations will not only drive innovation but also build trust, efficiency, and long-term sustainability in the evolving financial landscape.

Emerging Technologies

Emerging technologies are reshaping LLM operations, introducing new capabilities that address key challenges in the financial sector. These advancements enhance efficiency, scalability, and adaptability, allowing financial institutions to meet regulatory requirements and market demands more effectively. This section explores three critical areas driving innovation: quantum computing, model architecture advancements, and deployment strategies.

Quantum Computing and Finance

Quantum computing is transforming computational power, offering new possibilities for solving complex financial problems that classical computers struggle to address. Its integration into LLM operations (LLMOps) has the potential to drive breakthroughs in processing speed, analytical capabilities, and decision-making for financial institutions.

Note Quantum computing has the potential to revolutionize financial AI, but its integration with LLMs requires significant infrastructure investments. Organizations should stay informed on advancements while planning for a gradual adoption strategy.

How Quantum Computing Enhances LLMs

Quantum computers process information using quantum bits (qubits), which can exist in multiple states simultaneously. This enables massive parallel computations, allowing quantum systems to handle complex calculations at speeds far beyond classical computing. For LLMs in finance, quantum computing can help overcome computational bottlenecks in training and inference, particularly for large-scale models that require vast amounts of data processing.

For example, JPMorgan Chase has been actively exploring quantum computing to optimize portfolio strategies and risk analysis. Their pilot initiatives demonstrate how quantum-enhanced models could eventually support large-scale AI systems, including LLMs, in financial decision-making. They partnered with IBM to explore quantum computing for portfolio optimization and risk analysis. In their research, they used quantum algorithms to model complex financial systems, showcasing how quantum hardware could augment AI capabilities in financial services (Fingerhuth et al., 2021)[1].

These efforts illustrate how quantum-enhanced models can support LLMs with faster, more accurate, and scalable decision-making, especially for high-frequency trading, credit scoring, and stress testing.

Key Applications in Finance

As quantum computing continues to evolve, its impact on LLM-driven financial applications is becoming increasingly evident. Below are some of its most promising use cases:

- **Portfolio Optimization**: Quantum algorithms can rapidly analyze multiple investment variables and constraints, enabling more efficient portfolio selection and risk-adjusted investment strategies.

- **Faster Model Training**: Training LLMs for finance demands intensive computational power. Quantum computing can reduce training time, allowing for faster iterations, model updates, and improved accuracy.

- **Risk Analysis**: Quantum-enhanced simulations can model complex financial dependencies, leading to more precise risk assessments and scenario-based forecasting.

Integrating Quantum Computing with LLMOps

While quantum computing offers exciting possibilities, its integration into financial LLM operations comes with both challenges and opportunities. Some of the challenges are the following:

- **Hardware Limitations**: Current quantum systems lack stability and effective error correction, making large-scale adoption difficult.

- **Integration Complexity**: Financial institutions must reengineer existing LLMOps frameworks to leverage quantum computing effectively.

---

[1] *Fingerhuth, M., et al. (2021). Portfolio Optimization Using Quantum Algorithms. IBM Research & JPMorgan Chase. [Link to paper:* `https://arxiv.org/abs/1907.03044`*]*

- **High Costs**: Quantum infrastructure remains expensive, limiting its accessibility for mainstream financial applications.

Tip To prepare for the future of quantum-enhanced LLMs, financial institutions can start by experimenting with hybrid models, combining classical and quantum computing for incremental performance improvements.

While challenges exist, quantum computing presents significant opportunities for financial institutions looking to enhance LLM operations. Some of the opportunities are the following:

- **Efficiency Gains**: Quantum computing can accelerate model training and inference, leading to cost savings and performance improvements.

- **New Insights**: The ability to process complex datasets in parallel unlocks insights previously unattainable with classical methods.

Despite current limitations, quantum computing holds immense potential for transforming financial AI applications. As the technology matures, financial institutions that invest in quantum-compatible LLMOps strategies will be well-positioned to gain a competitive edge in high-speed analytics, predictive modeling, and risk assessment.

Chapter 8 Future Trends in LLM Ops for Finance

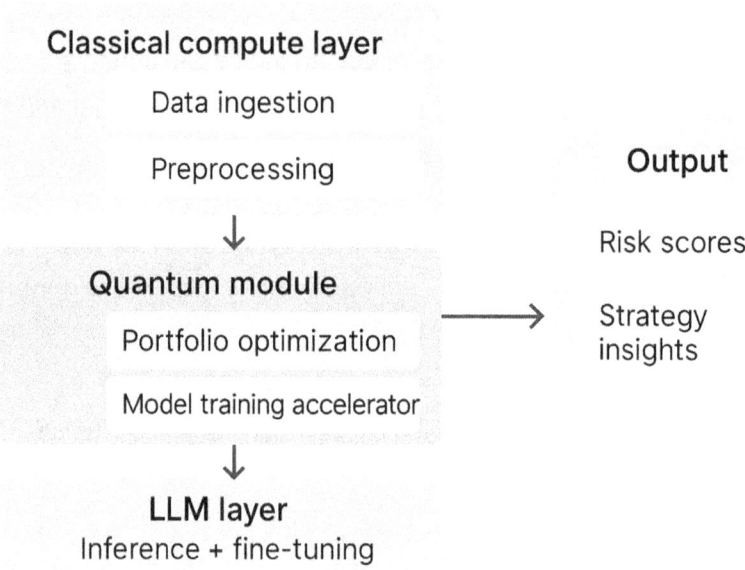

Figure 8-1. Quantum Enhanced LLM Ops Workflow

Figure 8-1 illustrates a **hybrid system integration architecture** combining classical computing, quantum modules, and LLM (large language model) layers to support advanced AI applications in financial services.

Key Components

1. **Classical Compute Layer**

    This is the entry point of the pipeline, responsible for:

    - **Data Ingestion:** Collecting structured and unstructured financial data (e.g., transactions, market feeds).

    - **Preprocessing:** Cleaning and formatting the data to ensure compatibility with downstream AI systems.

2. **Quantum Module**

   A specialized processing layer that performs computationally intensive tasks using quantum computing. It consists of

   - **Portfolio Optimization:** Solving complex financial optimization problems (e.g., asset allocation) using quantum algorithms.

   - **Model Training Accelerator:** Speeding up certain parts of LLM training or parameter search through quantum parallelism.

3. **LLM Layer**

   The LLM performs

   - **Inference and Fine-Tuning:** Applying the trained model to generate insights and adapting it to evolving data contexts in finance.

4. **Output Module**

   The final results delivered by the system:

   - **Risk Scores:** Probabilistic assessments of credit, market, or operational risk.

   - **Strategy Insights:** Actionable intelligence for decision-making, such as investment strategies or compliance recommendations.

This workflow exemplifies how **quantum computing can enhance LLMOps pipelines** by addressing bottlenecks in training and optimization. By tightly integrating quantum modules with classical AI systems, financial institutions can achieve faster training times and improved decision accuracy and gather deeper insights from complex datasets.

Caution Quantum computing is still in its early stages, with hardware limitations and high costs posing significant challenges. Rushing adoption without a clear roadmap may lead to inefficient investments and operational setbacks.

CHAPTER 8 FUTURE TRENDS IN LLM OPS FOR FINANCE

Advancements in Model Architectures

Modern large language models (LLMs) rely on transformer architectures, which form the foundation of their efficiency, scalability, and adaptability. As financial applications demand faster, more accurate, and resource-efficient AI models, continuous innovations in transformer architectures are driving improvements in performance, computational efficiency, and energy usage.

Innovations in Transformer Architectures

New developments in transformer design are making LLMs faster, more efficient, and better suited for complex financial tasks. Below are some key advancements:

- **Sparse Transformers**: By focusing computational power on the most relevant data, these models reduce processing time and memory usage. This is particularly beneficial for real-time transaction analysis in financial systems.

- **Low-Rank Adaptations**: Techniques like low-rank matrix factorization allow for more efficient fine-tuning, lowering computational costs for specialized applications such as fraud detection and credit scoring.

- **Energy-Efficient Models**: New transformer architectures are designed to minimize energy consumption during both training and inference, aligning with the financial sector's increasing focus on sustainability.

These advancements in transformer architectures are driving greater efficiency, scalability, and sustainability in financial AI applications. By adopting these innovations, financial institutions can enhance performance, reduce operational costs, and stay ahead in an increasingly data-driven landscape.

Tip When adopting specialized LLMs for finance, focus on models optimized for specific use cases like fraud detection, risk assessment, and compliance. Selecting the right architecture reduces costs and maximizes efficiency.

Specialized Models for Finance

To meet the unique demands of the financial industry, LLMs are being fine-tuned for specific use cases. Some of the most impactful specialized models include

- **Fraud Detection Models**: Tailored to detect fraudulent transactions by analyzing historical patterns and real-time financial data.

- **Credit Scoring Models**: Advanced transformers integrate multiple data sources to generate highly accurate credit risk assessments, improving lending decisions.

- **Regulatory Compliance Models**: Designed to parse legal and financial documents, these models assist institutions in ensuring adherence to evolving financial regulations.

By using specialized LLMs, financial institutions can enhance security, optimize risk assessment, and streamline regulatory compliance, ensuring more accurate decision-making and improved operational efficiency in an evolving financial landscape.

As shown in Table 8-2, LLM deployment strategies vary across cloud-native, hybrid, and edge modes. Each offers trade-offs in latency and security, making them suitable for batch tasks, balanced workloads, or real-time fraud detection.

Table 8-2. LLM Deployment Strategy Comparison

Deployment Mode	Latency	Security	Use Case Fit
Cloud-Native	Medium	Moderate	Scalable batch tasks
Hybrid	Low	High	Balanced workloads
Edge	Very Low	Very High	Real-time fraud detection

Caution Implementing new transformer architectures requires significant computational resources. Organizations must evaluate hardware capabilities, cost implications, and regulatory constraints before large-scale deployment.

Cutting-Edge Research and Tools

The latest research in financial AI has led to the development of highly specialized LLMs and tools that enhance financial decision-making and automation:

- **GPT-Fin**: A finance-specific transformer fine-tuned for tasks such as market sentiment analysis and trading signal predictions.

- **Graph-Enhanced Transformers**: These models incorporate financial knowledge graphs to improve the contextual understanding of market dynamics and risk factors.

- **OpenAI Codex for Finance**: A powerful tool that automates coding tasks in financial applications, including generating trading algorithms and risk models.

As transformer architectures continue to evolve, LLMs in finance will become even more precise, efficient, and adaptable. By adopting specialized models and taking advantage of the latest advancements in AI, financial institutions can enhance risk management, optimize decision-making, and ensure regulatory compliance in an increasingly complex market.

Note Transformer-based LLMs continue to evolve, with ongoing research in efficiency, energy consumption, and scalability. Financial institutions should regularly assess the latest developments to stay ahead in AI-driven decision-making.

Cloud-Native and Edge LLM Deployments

As financial institutions prioritize real-time data processing, security, and compliance, deployment strategies are shifting toward cloud-native and edge computing solutions. These approaches provide the necessary scalability, speed, and security to optimize LLM operations in finance.

Hybrid Deployments

To balance performance, security, and scalability, financial institutions are adopting hybrid deployment models that combine cloud-native and on-premises infrastructures.

- **Cloud-Native Solutions**: Cloud platforms offer scalable infrastructure for training and deploying LLMs, making them ideal for large-scale data processing, portfolio analysis, and customer service automation.

- **Hybrid Deployments**: Combining cloud and on-premises infrastructure allows institutions to process sensitive data locally while using the cloud for compute-intensive tasks like risk modeling and fraud detection.

- **Real-Time Processing**: Hybrid models ensure low-latency performance, making them suitable for time-sensitive applications such as market trend analysis, automated trading, and fraud detection.

By using hybrid deployments, financial institutions can achieve the optimal balance between scalability, security, and performance, ensuring real-time, efficient, and compliant AI-driven financial operations.

Tip Hybrid deployments provide the best of both worlds—the scalability of the cloud and the security of on-premises infrastructure. For compliance-sensitive financial applications, consider a hybrid model to balance performance and data privacy.

The Role of Edge Computing

Edge computing is redefining LLM deployments by processing data closer to the source, improving both response times and security in financial applications.

- **Reduced Latency**: Deploying LLMs at the edge (e.g., local servers, banking kiosks, or ATMs) enables instant transaction verification and real-time fraud prevention.

- **Enhanced Security**: Edge computing minimizes data exposure to external networks, reducing compliance risks and ensuring better data privacy.

- **Key Applications**: Financial institutions use edge computing for real-time fraud detection, mobile banking services, and automated teller systems, providing faster, more secure transactions.

To guide deployment decisions for large language models (LLMs) in financial services, it's important to match the technical requirements and compliance sensitivities of each use case with the most suitable infrastructure. Table 8-3 outlines common financial use cases and recommends whether **edge**, **hybrid**, or **cloud-based** deployments are most appropriate, based on factors like latency, data privacy, and regulatory complexity.

Table 8-3. Matching LLM Deployment Strategy to Financial Use Case

Use Case	Recommended Deployment	Reason
Fraud Detection	Edge	Requires low latency and immediate response at ATMs or POS terminals
AML/KYC Verification	Hybrid	Combines real-time checks with backend regulatory document analysis
Customer Support	Cloud/Hybrid	Less sensitive data; benefits from centralized learning and updates
Transaction Monitoring	Edge	On-device filtering of suspicious activity patterns
Regulatory Reporting	Cloud	High storage and archival needs; less urgency

Note Edge computing reduces dependency on centralized cloud servers, enabling low-latency financial transactions. This is especially beneficial for real-time fraud detection and mobile banking where quick decision-making is critical.

Emerging technologies are reshaping LLM operations in the financial sector:

- Quantum computing promises to revolutionize computational speed and efficiency.

- Transformer architecture advancements continue to improve model efficiency and domain-specific applications.

- Hybrid and edge deployments offer scalability, security, and flexibility, ensuring faster, safer, and more efficient AI-powered financial services.

By adopting these innovations, financial institutions can enhance operational efficiency, strengthen security, and unlock new AI-driven opportunities in the evolving financial landscape.

Caution While edge computing enhances security, infrastructure costs and maintenance can be significant. Financial institutions should carefully evaluate the trade-offs between latency improvements and deployment complexity before adopting widespread edge-based solutions.

Predictive Maintenance of Models

Predictive maintenance is essential for ensuring the sustained performance and reliability of large language models (LLMs) in operational settings. In the financial sector, where data streams are dynamic and the stakes are high, maintaining model accuracy and stability is critical. This section explores proactive monitoring techniques, automated updates, and strategies for ensuring long-term model reliability.

As shown in Table 8-4, predictive maintenance relies on performance, operational, and drift detection metrics. These evaluate output quality, runtime efficiency, and shifts in data distribution.

Table 8-4. Key Metrics for Predictive Maintenance

Metric Type	Examples	Purpose
Performance	Accuracy, Precision, Recall, F1-score	Evaluate model output quality
Operational	Latency, Throughput, Memory Usage	Measure runtime efficiency
Drift Detection	KL Divergence, Wasserstein Distance, PSI	Identify changes in data distribution

Proactive Monitoring Techniques

Ensuring the sustained performance and reliability of large language models (LLMs) is critical for financial applications, where real-time accuracy impacts fraud detection, credit risk assessment, and compliance. Proactive monitoring helps identify issues before they affect operations, allowing institutions to maintain high performance, regulatory compliance, and model integrity.

Importance of Real-Time Model Monitoring

To maintain consistent accuracy and operational efficiency, financial institutions must track real-time performance metrics, ensuring models adapt to evolving market conditions.

Caution Ignoring real-time monitoring can result in performance degradation, compliance violations, and financial losses. Institutions relying on static LLMs without adaptive learning strategies risk falling behind in an increasingly AI-driven financial landscape.

- **Detecting Model Drift**: Over time, data distributions may shift, causing a decline in model accuracy. Continuous monitoring detects drift early, preventing performance degradation. Modern MLOps platforms like Arize AI and Fiddler AI enable real-time drift detection, root cause analysis, and visualization of feature importance, making them ideal tools for maintaining model integrity in production.

- **Maintaining Operational Continuity**: In mission-critical applications like fraud detection and credit risk assessment, even small inaccuracies can lead to financial losses or compliance risks. Real-time tracking ensures stable model performance.

- **Compliance and Risk Mitigation**: Automated monitoring can detect biases, inaccuracies, or inconsistencies that may violate regulatory standards, helping organizations stay compliant with frameworks like Basel III and GDPR.

Real-time monitoring ensures LLMs remain accurate, compliant, and effective, safeguarding financial institutions against operational disruptions and regulatory risks.

> **Tip** Regularly retrain and fine-tune LLMs using the most recent financial data to maintain high accuracy and adapt to evolving risks. Automating retraining workflows can further enhance efficiency.

As shown in Table 8-5, regulations like GDPR, Basel III, the EU AI Act, and CCPA shape LLM deployment. Their implications span privacy rights, risk mitigation, ethical AI governance, and consumer data protections.

Table 8-5. LLM Deployment Strategy Comparison

Regulation	Focus Area	Implication
GDPR	Data Privacy	Consent management, right to explanation
Basel III	Risk Mitigation	Stress testing, model robustness
AI Act (EU)	Ethical AI Governance	Bias audits, transparency logs
CCPA	Consumer Data Rights	Right to opt-out, data sharing restrictions

Tools and Frameworks

To effectively monitor, diagnose, and respond to model drift, financial institutions leverage a variety of tools and frameworks:

- **Open-Source Tools**: Libraries like Evidently AI and Deepchecks provide interactive dashboards for tracking model performance and detecting drift.

- **Custom Monitoring Pipelines**: Organizations can build tailored monitoring solutions using tools like Apache Kafka for streaming financial data and TensorBoard for visualization.

- **Cloud-Native Solutions**: Platforms such as AWS SageMaker, Google Vertex AI, and Azure ML offer integrated monitoring and model drift detection for deployed LLMs.

By using the right monitoring tools, financial institutions can automate drift detection and quickly address performance degradation, ensuring models stay accurate and relevant.

CHAPTER 8 FUTURE TRENDS IN LLM OPS FOR FINANCE

	Data	Training	Inference	Output
GDPR	Privacy		Bias Mitigation	
Basel III		Risk Management		
AI Act			Bias Mitigation	
CCPA				Explainability

Figure 8-2. Regulatory Compliance Framework Map

The **Regulatory Compliance Framework Map** provides a clear visualization of how different global regulatory standards—such as GDPR, Basel III, the EU AI Act, and CCPA—apply across the lifecycle of LLM operations in finance. Structured as a matrix, it maps compliance focus areas to four key stages: data handling, model training, inference, and output. For instance, GDPR emphasizes privacy at the data stage and bias mitigation during inference, while Basel III focuses on risk management during model training. The AI Act reinforces bias mitigation in inference, and CCPA mandates explainability in model outputs. Figure 8-2 serves as a practical guide for aligning AI development with evolving legal obligations, helping financial institutions build responsible, auditable, and regulation-compliant AI systems. It highlights that compliance is not an afterthought but an integrated part of designing trustworthy LLM workflows in financial applications.

CHAPTER 8 FUTURE TRENDS IN LLM OPS FOR FINANCE

CASE STUDY: EARLY DETECTION OF MODEL UNDERPERFORMANCE IN FRAUD PREVENTION

A large financial institution deployed an LLM-powered fraud detection system to analyze transaction patterns. Over time, the system began missing fraudulent transactions, leading to increased financial losses.

Through real-time monitoring, analysts detected a drop in model accuracy caused by evolving fraud tactics. Proactive retraining using recent transaction data restored accuracy within hours, reducing financial losses by 15%.

This case highlights the importance of proactive model monitoring, enabling institutions to quickly detect performance declines and implement corrective actions.

Predictive maintenance strategies such as proactive monitoring, automated updates, and model retraining ensure that LLMs remain resilient, accurate, and compliant in dynamic financial environments. By investing in real-time tracking and drift detection, financial institutions can enhance decision-making, mitigate risks, and sustain long-term AI-driven success.

Note Even minor shifts in data distributions can significantly impact model accuracy. Implementing continuous monitoring solutions helps detect drift early and prevent costly errors.

Automated Updates and Retraining Pipelines

Automation is a critical component of predictive maintenance, allowing financial institutions to address model performance issues proactively without manual intervention. Automated retraining pipelines ensure that LLMs remain accurate, efficient, and aligned with evolving financial trends and risks.

Strategies for Automating Retraining

To maintain LLM accuracy and relevance, organizations implement automated retraining strategies that respond to performance fluctuations and data shifts.

- **Scheduled Retraining**: Models are updated at regular intervals using new financial data, ensuring they stay relevant and accurate.
- **Event-Triggered Retraining**: Significant performance drops or data drift automatically trigger retraining workflows, minimizing the risk of model degradation.
- **Active Learning**: The system selects the most informative data samples for retraining, optimizing resource usage while maintaining high model accuracy.

By integrating automated retraining strategies, financial institutions can ensure models evolve with real-world data, reducing drift and improving decision-making.

Role of Reinforcement Learning in Adaptive Models

Reinforcement learning (RL) plays a key role in adaptive LLMs, allowing them to dynamically adjust to new environments and continuously refine their outputs based on feedback.

- **Dynamic Adaptation**: RL enables models to self-improve by learning from new interactions, patterns, and market changes.
- **Application Example**: In customer service applications, reinforcement learning helps AI chatbots refine responses by analyzing user interactions and feedback, improving customer satisfaction over time.

Reinforcement learning enhances model adaptability, enabling LLMs to stay responsive and relevant in fast-changing financial applications.

Benefits of Automated Workflows

Automating retraining pipelines provides multiple operational advantages, ensuring LLMs remain scalable, efficient, and error-free.

- **Consistency**: Automation reduces human errors, ensuring timely and standardized updates.

- **Scalability**: Automated workflows support large-scale deployments, allowing institutions to update multiple models across financial services simultaneously.

- **Cost Efficiency**: Reducing manual intervention lowers operational costs while maintaining high model accuracy.

Automated workflows provide financial institutions with a scalable, cost-effective approach to maintaining AI-driven decision-making accuracy.

Ensuring Long-Term Model Reliability

For LLM operations to succeed in the long run, institutions must focus on model reliability, performance metrics, and efficient update strategies. A well-defined maintenance framework ensures that models remain stable, effective, and resilient in financial applications.

Metrics for Measuring Model Health and Stability

Tracking the right performance indicators helps financial institutions assess model reliability and take corrective action when needed.

- **Performance Metrics**: Accuracy, precision, recall, and F1-score measure model effectiveness in decision-making.

- **Operational Metrics**: Latency, throughput, and resource utilization provide insights into model efficiency and responsiveness.

- **Drift Metrics**: Statistical measures like KL divergence and Wasserstein distance help detect shifts in data distributions that could impact predictions.

By continuously monitoring key metrics, financial institutions can detect issues early and ensure models remain accurate and efficient.

Approaches to Minimize Downtime

Updating models without disrupting financial operations is crucial. Several strategies help institutions implement seamless updates while ensuring stability.

- **A/B Testing**: Running new and existing models side by side allows teams to evaluate performance without disrupting services.

- **Canary Releases**: Rolling out updates to a small subset of users first ensures stability before full-scale deployment.

- **Shadow Testing**: Deploying updated models in parallel with live systems validates performance in real-world conditions before full activation.

Minimizing downtime ensures financial models remain operational, preventing service disruptions and performance failures.

Examples of Successful Implementations

Financial institutions are already leveraging predictive maintenance strategies to improve model accuracy, risk assessment, and customer engagement.

- **Fraud Detection Systems**: A leading bank reduced fraud losses by 20% using automated retraining pipelines, which updated the fraud detection model weekly.

- **Credit Scoring Models**: A financial institution improved credit risk predictions by 30% by integrating real-time monitoring and automated retraining workflows.

- **Customer Support Chatbots**: A banking AI chatbot enhanced customer satisfaction rates by continuously refining its responses through reinforcement learning.

These real-world examples highlight the impact of predictive maintenance in driving efficiency, improving accuracy, and reducing financial risks.

Predictive maintenance is a cornerstone of sustainable AI operations in finance. By combining proactive monitoring, automated updates, and long-term reliability strategies, financial institutions can maintain high-performance LLMs while optimizing costs and ensuring compliance. As financial AI systems continue to evolve, investing in predictive maintenance will be key to staying competitive and resilient in a rapidly changing market.

CHAPTER 8 FUTURE TRENDS IN LLM OPS FOR FINANCE

The Future of AI in Finance

As large language models (LLMs) become central to the financial sector, their future will be shaped by regulatory advancements, emerging applications, and collaborative ecosystems. Financial institutions must stay ahead of compliance requirements, ethical considerations, and new AI-driven opportunities to remain competitive.

As shown in Figure 8-3, the financial AI ecosystem thrives on collaboration between financial institutions, AI providers, academia, and the open-source community. This synergy drives innovation, transparency, and practical adoption.

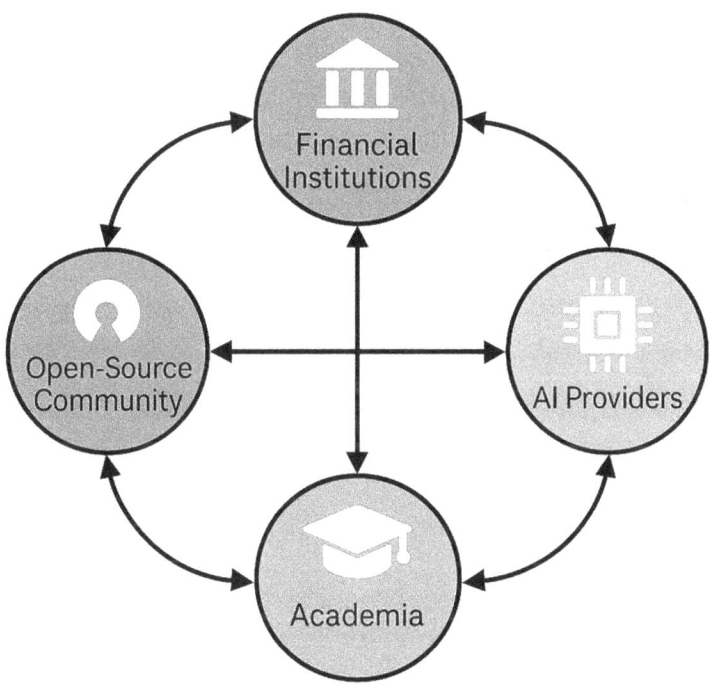

Figure 8-3. *Collaborative Ecosystem in Financial AI*

Regulatory Standards and Ethical Considerations

The increasing reliance on LLMs in finance comes with greater regulatory scrutiny. As AI adoption grows, governments and financial regulatory bodies will introduce new frameworks to ensure fair, ethical, and transparent AI operations. Institutions must develop strong governance strategies to remain compliant while maximizing AI's benefits.

CHAPTER 8   FUTURE TRENDS IN LLM OPS FOR FINANCE

Anticipated Changes in Regulatory Frameworks

Regulatory frameworks for financial AI are expected to become more dynamic, sector-specific, and globally aligned.

- **Global Coordination**: Organizations such as the Financial Stability Board (FSB)[2]. and the International Organization of Securities Commissions (IOSCO)[3] are likely to introduce standardized AI governance policies.

- **Sector-Specific Rules**: LLMs will face stricter oversight in financial applications, ensuring compliance with fairness, security, and transparency standards:

  - Credit scoring models must prove fairness and avoid discriminatory biases.

  - Trading algorithms will undergo stricter evaluations to prevent market manipulation.

- **Dynamic Regulation**: Regulators will continuously evolve compliance requirements to align with new AI capabilities and risks.

Staying ahead of evolving regulatory standards is essential for financial institutions to ensure compliance, build trust, and prevent legal risks.

Ensuring Compliance

Financial institutions must ensure that LLMs adhere to global data protection laws while maintaining ethical AI principles.

---

[2] *Financial Stability Board (2022). AI and Machine Learning in Financial Services: Regulation and Supervisory Implications.*
[3] *IOSCO (2021). The Use of Artificial Intelligence and Machine Learning by Market Intermediaries and Asset Managers.*

- **Data Privacy**: Compliance with GDPR, CCPA, and evolving data laws requires strict data handling policies, ensuring LLMs do not expose personal information.

- **Bias Mitigation**: AI models must undergo continuous bias audits to ensure fair decision-making across diverse demographics.

- **Ethical AI Guidelines**: Institutions should align with ethical AI frameworks emphasizing transparency, accountability, and user consent.

By integrating ethical AI principles, financial organizations can ensure responsible AI use, fostering trust among regulators, customers, and stakeholders.

Impact of Explainability and Transparency

Regulatory bodies emphasize the need for explainable AI (XAI) in financial applications, particularly for credit scoring, risk assessment, and fraud detection.

- **Explainable AI (XAI)**: Financial institutions must develop models that provide clear, understandable explanations for their decisions.

- **Model Audits**: Comprehensive documentation and audit trails will be essential for demonstrating regulatory compliance.

- **Transparency in Decision-Making**: LLMs must justify their predictions and recommendations, ensuring clarity for financial regulators and customers.

Explainability and transparency are key to regulatory compliance and trust-building, ensuring LLMs remain ethical and accountable in financial decision-making.

Caution Failure to establish clear governance structures in collaborative AI projects can lead to compliance risks, intellectual property disputes, and security vulnerabilities. Financial institutions must define clear data-sharing policies and ethical guidelines when co-developing LLMs.

CHAPTER 8 FUTURE TRENDS IN LLM OPS FOR FINANCE

Emerging Applications of LLMs in Finance

LLMs are transforming financial services by handling complex tasks, improving market insights, and providing hyper-personalized solutions. As AI capabilities expand, institutions can leverage LLMs for deeper market analysis, customized financial advisory, and advanced risk assessment.

New Use Cases

The next generation of financial AI applications will leverage LLMs for personalized advisory and multimodal analysis.

- **Hyper-Personalized Financial Advisory**: LLMs equipped with customer-specific data can generate highly customized investment and risk management advice.

    - Example: A virtual financial advisor could design customized retirement plans based on a user's income, expenses, and long-term goals.

- **Multimodal Analysis**: LLMs will integrate text, numerical, and visual data for deeper financial insights.

    - Example: Combining stock market news sentiment with trading signals to enhance investment strategies.

By integrating multimodal capabilities, LLMs will unlock new dimensions of financial analysis, enhancing customer engagement and investment decisions.

As shown in Table 8-6, emerging technologies like quantum computing, sparse transformers, hybrid cloud, edge computing, and reinforcement learning offer powerful benefits. However, each faces challenges such as cost, complexity, or infrastructure overhead.

Table 8-6. Comparative Overview of Emerging Technologies

Technology	Benefit	Challenge
Quantum Computing	Massive parallel computation for model training and risk analysis	Hardware immaturity, high costs
Sparse Transformers	Reduces compute load and latency in financial inference tasks	Complex tuning, may reduce model generalization
Hybrid Cloud	Scalable and compliant infrastructure for sensitive workloads	Integration complexity with legacy systems
Edge Computing	Enables real-time, secure, on-device inference	Maintenance and infrastructure overhead
Reinforcement Learning	Continuously adapts LLMs based on feedback and performance	Requires high-quality feedback loops and tuning

Potential for LLMs in Applications

LLMs will revolutionize financial forecasting and risk assessment by analyzing live data streams and running complex scenario simulations.

- **Real-Time Market Analysis**: AI models will process stock prices, news feeds, and social media trends to identify market movements and emerging risks.

- **Scenario Planning**: LLMs can simulate financial risks under various conditions, helping institutions prepare for regulatory changes, economic shifts, and crises.

  - Example: Predicting the impact of new regulations on loan portfolios and adjusting financial strategies accordingly.

Real-time market analysis and scenario planning will enable institutions to respond faster to market shifts, reducing risks and maximizing opportunities.

Example: Predicting Systemic Risks Through LLM-Driven Insights

A leading banking network deployed an LLM-powered risk analysis system to analyze global financial data and detect systemic threats.

- The model identified early warning signs of liquidity crises and market instabilities.
- By acting on AI-driven insights, the institution mitigated financial risks and prevented potential losses.

This example highlights the power of LLMs in systemic risk detection, helping financial institutions make proactive, data-driven decisions.

The future of LLMs in finance will be shaped by regulatory evolution, ethical considerations, and AI-driven innovation. Institutions that embrace explainability, compliance, and cutting-edge AI applications will gain a competitive advantage in risk management, market analysis, and financial services.

By staying ahead of regulatory changes and using AI for hyper-personalized insights and real-time decision-making, financial organizations can unlock new opportunities while ensuring transparency and accountability.

Collaboration Between Institutions and Technology Providers

The success of financial LLMs increasingly depends on strategic partnerships and community-driven innovations. Financial institutions are collaborating with AI providers, academia, and open-source communities to accelerate LLM development, deployment, and governance. These collaborations foster innovation, cost efficiency, and industry standardization, ensuring that LLMs remain effective and adaptable in the rapidly evolving financial sector.

Building Partnerships to Co-Develop Financial LLMs

Financial institutions are forming strategic alliances with AI providers and academic institutions to drive domain-specific advancements in LLMs.

- **Financial Institutions and AI Providers:** Banks, insurance companies, and fintech firms are working with AI technology providers to develop specialized financial LLMs.

- **Example:** A leading bank partnered with an AI firm to create a compliance-focused LLM capable of parsing complex regulatory documents, reducing manual workload.

- **Academia-Industry Partnerships**: Universities and research labs collaborate with financial organizations to explore cutting-edge LLM applications in risk assessment, fraud detection, and financial forecasting.

Collaborative partnerships accelerate financial LLM advancements, ensuring models remain innovative, scalable, and aligned with industry-specific needs.

The Role of Open-Source Contributions

Open-source platforms are driving LLM innovation, allowing financial institutions to benefit from shared research, tools, and pre-trained models.

- **Community Contributions:** Platforms like Hugging Face, TensorFlow, and PyTorch provide open-source LLM frameworks, enabling faster experimentation and deployment.

- **Knowledge Sharing:** Financial institutions contribute datasets, benchmarks, and best practices to the open-source ecosystem, enhancing financial AI applications.

- **Cost-Effective Development:** Open-source tools allow smaller financial institutions to adopt LLMs at lower costs, reducing the need for expensive proprietary models.

Leveraging open-source contributions enhances LLM accessibility, fosters innovation, and accelerates financial AI development.

Tip Financial institutions should actively engage with open-source AI communities, contributing best practices and financial datasets to help improve LLM development and governance.

CHAPTER 8 FUTURE TRENDS IN LLM OPS FOR FINANCE

Future Prospects for Collaborative Ecosystems in Financial AI

The future of financial AI development will be shaped by collaborative ecosystems that promote standardization, innovation, and global partnerships.

- **Standardization**: Collaboration between financial institutions, AI firms, and regulatory bodies will lead to industry-wide LLM governance frameworks.

- **Innovation Hubs**: Financial institutions and technology providers will co-develop solutions in shared innovation labs, accelerating AI advancements.

- **Cross-Border Initiatives**: International collaborations will drive the creation of globally applicable LLMs, addressing financial risks, compliance, and regulatory challenges across jurisdictions.

Collaborative ecosystems will drive the next wave of financial AI innovation, ensuring standardized, scalable, and globally relevant LLM solutions.

Note Strategic partnerships with AI providers and academia can accelerate LLM innovation, enabling financial firms to develop customized, high-performance models tailored to industry-specific needs.

The future of AI in finance will be defined by alignment with regulatory standards, expansion into new applications, and collaborative development. Financial institutions that embrace compliance, innovation, and partnerships will be better positioned to navigate evolving LLM operations and maintain a competitive edge.

A forward-looking approach that integrates regulatory alignment, technological advancements, and collaborative ecosystems will foster resilience, efficiency, and growth in the financial AI landscape.

Conclusion

As large language models (LLMs) reshape the financial sector, their success will depend on regulatory alignment, ethical AI adoption, advanced deployment strategies, and collaborative innovation. Financial institutions must navigate evolving compliance frameworks, integrate emerging AI applications, and leverage strategic partnerships to stay competitive in an increasingly AI-driven landscape.

The future of LLM operations in finance will be shaped by real-time adaptability, transparency, and responsible AI governance. Institutions that proactively implement predictive maintenance, optimize deployment models, and collaborate with AI leaders will position themselves at the forefront of financial innovation and resilience.

The financial industry is undergoing a transformative shift, with LLMs playing an increasingly central role in risk management, fraud detection, and personalized financial services. To fully realize AI's potential, financial institutions must strategically invest in regulatory compliance, model optimization, and industry-wide partnerships.

Key Points

- **Regulatory Evolution:** Financial AI will be governed by dynamic, globally coordinated regulations requiring greater transparency and explainability in model decisions.

- **Ethical AI Practices:** Institutions must prioritize fairness, bias mitigation, and data privacy to ensure trustworthy AI adoption in financial applications.

- **Explainability and Compliance:** Explainable AI (XAI) and audit trails will be mandatory components of future financial LLMs to meet regulatory and ethical standards.

- **Advancements in Transformer Architectures:** Sparse transformers, low-rank adaptations, and energy-efficient models will drive higher efficiency and scalability in LLMs.

- **Hybrid and Edge Deployments:** Combining cloud-native, hybrid, and edge AI solutions will ensure low-latency, secure, and high-performance financial applications.

- **Predictive Maintenance:** Automated monitoring and event-triggered retraining will be essential to prevent model drift and maintain accuracy in financial LLMs.

- **Emerging AI Applications:** LLMs will power hyper-personalized financial advisory, multimodal analysis, and real-time market simulations for advanced financial insights.

- **Collaborative Ecosystems:** Financial institutions, AI technology providers, and open-source communities will drive innovation, standardization, and cost-effective AI solutions.

- **Cross-Border AI Collaboration:** Global partnerships will accelerate financial AI research and standardization, ensuring regulatory harmonization and risk management.

- **Future-Proofing Financial AI:** Institutions that embrace regulatory agility, real-time AI insights, and collaborative AI development will lead the next generation of financial intelligence.

Index

A

Accuracy, 207
Active learning, 219, 220
Adaptive Windowing (ADWIN), 204
Airflow, 42
Algorithmic trading, 12, 179
Amazon Web Services (AWS), 42
Anomaly detection, 118–120
Anonymization, 139
Anti-money laundering (AML), 22, 143, 161, 181
Apache Kafka, 174, 183
Apache Spark, 41
Application programming interfaces (API), 169
    Apache Kafka, 174
    consumers, 177
    continuous updates, 171, 172
    data sources, 173
    deployment, 196
    latency, 170, 171
    microservices, 174, 175
    preprocessing layer, 174
    RESTful APIs, 169
    security, 170
    step-by-step process flow, 175–177
    storage, 178, 179
Artificial intelligence (AI), 1, 198
    accelerators, 54
    financial sector, 137
    regulatory standards, 266–270

Auditability, 158
Autoencoders, 69
Automated analysis, 9, 10
Automated testing, 108–110
Automating compliance, 8, 198
Automation
    benefits, 221, 262
    combined drift, 224
    compliance reporting, 195–198
    components, 221
    concept drift, 224
    data drift, 223, 224
    performance monitoring, 225
    real-time drift detection tools, 225
    regular updates, 222, 223
    reinforcement learning (RL), 262
    statistical tests, 224
    strategies, 262
AWS CloudWatch, 211, 212

B

Backward compatibility, 172, 173

C

California Consumer Privacy Act (CCPA), 152, 158, 162
Canary deployments, 110–113
Causal Language Modeling (CLM), 78
Central processing units (CPUs), 31, 117
Chatbots, 186–191

INDEX

Checkpointing, 81
Cloud infrastructure, 42, 43
Comparative analysis, 46
Compliance, 51–53, 136, 137, 139, 141, 144, 148, 204, 266, 267
    California Consumer Privacy Act (CCPA), 152
    General Data Protection Regulation (GDPR), 152
    Payment Card Industry Data Security Standard (PCI DSS), 153
Content Delivery Networks (CDNs), 124
Continuous integration/continuous deployment (CI/CD)
    automated testing, 108–110
    financial systems, 105
    LLM deployment, 105
    multiple versions, 107, 108
    testing phases, 106
Cost management, 44, 119
Cost savings, 197
COVID-19 pandemic, 216
Credit risk assessment, 10, 57, 61, 101, 179
Credit scoring, 70, 71, 104, 155, 159, 162, 180, 206
Customer service Chatbots, 186–191
Customization, 45
Custom monitoring, 212
Cutting-edge research, 254

D

Data access controls, 146
Data anonymization, 138, 140
    privacy, 142
Data augmentation, 63–66, 75, 76
Data cleaning, 62, 74, 75, 176
Data collection, 61, 73, 74, 152
Data governance, 76, 77
Data masking, 145
Data preparation, 91, 143, 144
    cleaning, 74, 75
    collection, 73, 74
    pipeline, 72
    splitting, 76
    tools and frameworks, 77
Data privacy and security, 230, 231
    implications, 163, 164
    model drift detection, 231, 232
    objectives, 136, 137
    operational effectiveness, 137–140
    structure, 135, 136
Data sensitivity, 20, 21, 46, 89, 128
Data storage, 145–148
    best practices, 149, 150
    cloud-based, 148, 149
    fraudulent activities, 150, 151
Data version control (DVC), 107
Deployment strategies
    monitoring and logging, 113–119
    performance and reliability, 104
    pipelines, 104–113
    structured approach, 104
Deployment tools, 39–41, 203
Docker, 39, 40
Documentation, 76, 77, 155
Domain-adaptive pretraining (DAPT), 68, 84, 93, 98–103
Domain shifts, 88
Drift analysis
    covariate shift detection, 226
    feature importance analysis, 225
    mitigating model drift, 226–228
    predictive monitoring, 226
Drift Detection Method (DDM), 204

E

Edge computing, 255–257
Emerging technologies
    cutting-edge research, 254
        model architectures, 252–254
        quantum computing, 247–251
Encryption, 48, 145, 146, 160, 161
End-user systems, 184
Equal Credit Opportunity Act (ECOA), 99
Error tracking, 115, 116
Explainable AI (XAI) techniques, 22, 70

F

Face transformers, 37
Federated learning, 161
Finance, LLMs, 72, 93, 104, 109, 115, 118, 128, 244
    accuracy and human error, 16
    analytical tools, 17, 18
    in applications, 269
    challenges
        bias, 22–24
        computational requirements, 27, 28
        data sensitivity and privacy, 20, 21
        data volatility, 25, 26
        drift, 25, 26
        fairness, 22–24
        interpretability and transparency, 21, 22
        large language models (LLMs) deployments, 26–28
    client experience, 18, 19
    collaborative ecosystems, 272
    compliance, 151–153
        best practices, 160, 161
    data complexity, 245
    efficiency and productivity, 15–19
    ethical considerations, 246
    financial institutions, 270, 271
    navigation, 246
    new use cases, 268, 269
    non-compliance, 154–157
        audit, 155
        documentation, 155
    open-source platforms, 271
    personalization, 18, 19
    predicting systemic risks, 270
    real-time risk management, 18
    regulatory compliance, 19
    technological advancements, 246
    use cases, 245
Financial forecasting, 12, 68
Financial-specific infrastructure, 53, 54
Financial Stability Board (FSB), 266
Fine-tuning, 93, 106, 107, 123, 129, 145, 188
    active, 87
    adaptive strategies, 69–73
    challenges
        computational, 90
        data-related, 89
        domain-specific challenges, 90
    compliance and ethics, 95, 96
    domain-adaptive pretraining (DAPT), 84
    domain shifts, 88
    evaluation and monitoring, 88, 95, 96
    few-shot and zero-shot learning, 85, 86
    high-quality data, 60–64
    hyperparameter tuning, 86
    life-cycle, 59
    multi-task, 87
    regularization, 86
    task-specific, 83, 84
    techniques, 67, 68

INDEX

Fine-tuning (*cont.*)
    training methodologies, 65–67
    transfer learning leverages, 85
Foundation model, 128
Fraud detection, 10, 48, 69, 70, 104, 113, 123, 129, 138, 142, 144, 145, 162, 179, 181, 185, 186, 206, 228
Fraud prevention, 260
Fraudulent activity, 14, 69, 129, 150, 151, 179

G

General Data Protection Regulation (GDPR), 152, 158
Generative Adversarial Networks (GANs), 63, 75
Global adoption, 6
Google Cloud Platform (GCP), 42
Google Vertex AI, 211
GPTQ (Post-Training Quantization), 95
Graphics processing units (GPUs), 30, 117

H

Hard disk drives (HDDs), 32
Hardware requirements
    central processing units (CPUs), 31
    graphics processing units (GPUs), 30
    memory configurations, 32, 33
    optimization, 33–35
    storage solutions, 31, 32
    tensor processing units (TPUs), 30, 31
High-demand applications
    case studies, 186–198
Homomorphic encryption, 160, 161
Hybrid approach, 47

Hybrid deployments, 255
Hyperparameters, 66
    tuning, 86, 107

I

Infrastructure Setup
    cloud *vs.* on-premises solutions, 42, 43
    compliance, 51–53
    hardware requirements, 29–35
    security guidelines, 48–50
    software stack, 35–42
In-memory storage systems, 126
International Organization of Securities Commissions (IOSCO), 266
Internet service providers (ISPs), 51
Interpretability, 70, 99
Iterative updates, 68

J

JPMorgan Chase, 247

K

Kubernetes, 40

L

Large language models (LLMs), 113
    advancements, 4
    application, 129, 130
    architecture, 4
    artificial intelligence, 1
    automated analysis, 9, 10
    capabilities, 5, 6
    compliance monitoring, 11
    customer service, 12, 13
    deployment strategies, 101

INDEX

efficiency gains, 7
financial forecasting, 12
financial institutions, 15–19
financial systems, 186–198
hybrid AI approaches, 13, 14
infrastructure Setup, 29–55
iterative improvement, 131, 132
maintenance, 238, 239
market growth, 6
monitoring and maintaining, 232–235
    adaptation, 203, 204
    challenges, 205
    challenges and best practices, 229
    long-term performance, 203
    performance metrics, 206–208
    reliability and efficiency, 205
predictive maintenance, 257–264
production and monitoring, 131, 132
risk assessment, 1
risk management, 10–12
simplified breakdown, 3
strategic benefits, 8, 9
streamline analysis, 2
training, 92
validation and evaluation, 130
Local Interpretable Model-Agnostic Explanations (LIME), 24
Logging, 81, 82, 103, 106, 113–119
Long-term model reliability
    approaches, 264
    health and stability, 263
    implementations, 264

M

Market forecasting, 12, 131, 132
Masked Language Modeling (MLM), 78
Message broker, 183

Microservices-based architecture, 175
Microsoft Azure, 42
Mitigating model drift
    adaptive learning techniques, 227
    Ensemble models, 228
    full retraining, 227
    incremental retraining, 227
    monitoring and alerts, 226
MLflow, 37, 107
MLOps, 38
Model compression, 121
Model drift, 215, 217, 223
Multi-factor authentication (MFA), 49, 146
Multi-task fine-tuning, 87

N

Natural language processing (NLP), 9, 31
Network attached storage (NAS), 32
Network redundancy, 51
Network security, 49
Non-compliance
    AI model integrity, 156, 157
    data minimization, 155
NVMe storage, 35

O

On-premises solutions, 44, 45
On-the-fly data processing, 183
Operational risk management, 11

P

Payment Card Industry Data Security Standard (PCI DSS), 153, 158
Performance metrics, 114, 115
    accuracy, 207
    domain-specific metrics, 209, 210

279

INDEX

Performance metrics (*cont.*)
    F1-score, 208
    latency, 208, 209
    proactive management, 206
    real-time monitoring, 210–212
    resource utilization, 209
Performance monitoring
    active learning, 219, 220
    adaptation, 215–217
    automating retraining, 220–223
    correlate metrics, 214, 215
    implementation alerts, 213
    key metrics, 213
    regulatory changes, 217
    retraining, 217
    review metrics, 213
Performance optimization
    scaling
        caching & load balancing, 125–127
        horizontal, 120
        latency reduction, 123–125
        life cycle, 127–132
        model compression & quantization, 121, 122
        practices, 121
        quantization, 122
        vertical, 120
Personal finance assistance, 12, 13
Personalization, 18, 19
Personally identifiable information (PII), 20, 135, 137
Portfolio optimization, 71, 72
Predictive maintenance, 235, 238
    automated updates, 261–263
    key metrics, 257
    proactive monitoring, 258–261
Privacy, 20, 21

Proactive monitoring
    real-time model, 258, 259
    tools and frameworks, 259
Prometheus, 210
Pruning, 122
Python library, 41
PyTorch, 36, 54

Q

Quantization, 122
Quantized Low-Rank Adaptation (Q-LoRA), 95
Quantum computing
    with LLMOps, 248–250
    quantum-enhanced models, 248
    use cases, 248

R

Real-time data processing
    architectural considerations, 181
    in finance, 179
    scalability strategies, 182–184
    techniques, 180
Real-time inference, 184
Real-time metric
    evidently AI, 211
    prometheus, 210
Recurrent neural networks (RNNs), 3
Regulatory Compliance Framework Map, 260
Regulatory enforcement, 162
Regulatory standards, AI
    anticipated changes, 266
    explainability and transparency, 267, 268
Reinforcement learning (RL), 237, 262

Resource monitoring, 117, 118
RESTful APIs, 175
Retraining, 217, 261–263
Risk management, 10–12, 71, 104
Role-based access control (RBAC), 49

S

Scalability, 221
Self-healing systems, 236, 237
Sentiment analysis, 8, 10, 36, 38, 43, 55, 73, 84, 191–195, 206
Software stack
    data processing, 41, 42
    deployment tools, 39–41
    integration tools, 41
    libraries and frameworks, 35–37
    model management tools, 37–39
Solid-state drives (SSDs), 31, 34
Source systems, 183
Specialized models, 253
Storage methods, 146
Strategic benefits, 8, 9
Synthetic data generation, 141
Synthetic Minority Over-sampling Technique (SMOTE), 63, 75, 91

T

Task-specific adaptation, 67
TensorBoard, 38
TensorFlow, 36, 40
TensorFlow Extended (TFX), 39
Tensor processing units (TPUs), 30
Tokenization, 148
Total Cost of Ownership (TCO), 46
Training, 138–142, 144, 148, 152, 156, 158, 159, 161, 187, 188, 190
    best practices, 92
    challenges, 82, 83
    checkpointing, 81
    distributed training, 79
    end-to-end lifecycle, 58
    vs. fine-tuning, 59, 60
    handling data, 81
    high-quality data, 60–64
    optimization algorithms, 80
    pretraining, 78, 79
    regularization, 80, 81
Transformer architectures, 252

U

Use Case, 210

V

Variational Autoencoders (VAEs), 63, 69, 75

W, X, Y, Z

Weights & biases (W&B), 38, 107

GPSR Compliance

The European Union's (EU) General Product Safety Regulation (GPSR) is a set of rules that requires consumer products to be safe and our obligations to ensure this.

If you have any concerns about our products, you can contact us on

ProductSafety@springernature.com

In case Publisher is established outside the EU, the EU authorized representative is:

Springer Nature Customer Service Center GmbH
Europaplatz 3
69115 Heidelberg, Germany